Studies in Computational Intelligence

Volume 1003

The series "Studies in Computational Intelligence" (SCI) publishes new developments and advances in the various areas of computational intelligence—quickly and with a high quality. The intent is to cover the theory, applications, and design methods of computational intelligence, as embedded in the fields of engineering, computer science, physics and life sciences, as well as the methodologies behind them. The series contains monographs, lecture notes and edited volumes in computational intelligence spanning the areas of neural networks, connectionist systems, genetic algorithms, evolutionary computation, artificial intelligence, cellular automata, self-organizing systems, soft computing, fuzzy systems, and hybrid intelligent systems. Of particular value to both the contributors and the readership are the short publication timeframe and the world-wide distribution, which enable both wide and rapid dissemination of research output.

Indexed by SCOPUS, DBLP, WTI Frankfurt eG, zbMATH, SCImago.

All books published in the series are submitted for consideration in Web of Science.

More information about this series at https://link.springer.com/bookseries/7092

Roger Lee
Editor

Computer and Information Science 2021 - Fall

Springer

Editor
Roger Lee
Software Engineering and Information
Technology Institute
Central Michigan University
Mount Pleasant, MI, USA

ISSN 1860-949X ISSN 1860-9503 (electronic)
Studies in Computational Intelligence
ISBN 978-3-030-90527-9 ISBN 978-3-030-90528-6 (eBook)
https://doi.org/10.1007/978-3-030-90528-6

This Springer imprint is published by the registered company Springer Nature Switzerland AG
The registered company address is: Gewerbestrasse 11, 6330 Cham, Switzerland

Preface

The purpose of the 21st IEEE/ACIS International Fall Virtual Conference on Computer and Information Science (ICIS 2021-Fall) held on October 13–15, 2021, in Xi'an, China, was to bring together researchers, scientists, engineers, industry practitioners, and students to discuss, encourage, and exchange new ideas, research results, and experiences on all aspects of computer and information science, and to discuss the practical challenges encountered along the way and the solutions adopted to solve them. The conference organizers have selected the best 12 papers from those papers accepted for presentation at the conference in order to publish them in this volume. The papers were chosen based on review scores submitted by members of the program committee and underwent further rigorous rounds of review.

In Chapter 1, Yapeng Wang, Ruize Jia, Chan Tong Lam, Ka Cheng Choi, Koon Kei Ng, Xu Yang, and Sio Kei Im present research and implementation of an automatic speech recognition (ASR) engine for Portuguese language.

In Chapter 2, Wissem Abbes, Zied Kechaou, Amir Hussain, and Adel M. Alimi propose a binary particle swarm optimization (BPSO)-based approach that is useful for maintaining an effective service bursting optimization within the hybrid cloud framework.

In Chapter 3, Keontaek Oh and DaeSoo Kim identify an empirically ideal profile of the cash conversion cycle (CCC) and its elements of the top performance group in terms of the average net sales growth rate during the global financial crisis and recovery periods, using the data of the Forbes Global 2000 ranking firms in the automotive industry.

In Chapter 4, Changhwa Baek and Eunil Son investigate the concept and characteristics of the 6th industry that is leading new changes in rural areas.

In Chapter 5, Sujeong Baek, Dong Oh Kim, Seo Jin Lee, Na Hyeon Yu, and Su In Chea identify relevant parameters for the real-time condition of a vacuum gripper system to detect the corresponding operation status. When using a measurement system, they conducted an experiment to discover effects of control parameters to pickup operation of a vacuum gripper, whether a gripper succeeds or fails in holding a part.

In Chapter 6, Youn Sung Kim, Seungbeom Kim, Kyungmi Bae, and Minseo Park present a data envelopment analysis (DEA), a widely accepted analysis model for measuring efficiency. The study derived appropriate factors for efficiency assessment in digital service of university library by the literature review.

In Chapter 7, Seo Young Kim, Tae Hee Kim, Youn Sung Kim, and Min Seo Park analyze changes in airport service environment and process, to derive improvements on pain point found through in-depth interviews and persona production of employees and customers, and to establish an improved airport service process and service environment by reflecting the needs of customers.

In Chapter 8, Jin-Hee Lee conducted an investigation on how the empirical value of untact marketing affects consumer satisfaction and consumer's willingness to repurchase, centered on service mobile application. This investigation intends to contribute to the establishment of untact marketing strategy.

In Chapter 9, Jung Jae Lee finds and utilizes stakeholders' active involvement in the activation of digital copyright exchanges through the development of blockchain-based copyright contracts and distribution platform technologies that provide a total service from copyright contracts to distribution by developing high-capacity content data storage and retrieval technologies.

In Chapter 10, Yoo Jeong Lee, Ji Woo Choi, Hyun Woo Nam, and Sae Young Shin present a strategic alternative to the development of the smart beauty industry related to personalized beauty products based on big data on skin information tailored to consumer needs.

In Chapter 11, Wonhyung Park and Hoo-Ki Lee examine countermeasures against drone communication (Wi-Fi, Bluetooth, and GPS) vulnerabilities.

In Chapter 12, Ioannis Daramouskas, Dimitrios Mitroulias, Isidoros Perikos, Michael Paraskevas, and Vaggelis Kapoulas examine the localization capabilities of LoRa networks in terms of the localization error. Various experiments were performed to assess the localization capabilities of LoRa technology in the real-world setup.

It is our sincere hope that this volume provides stimulation and inspiration, and that it will be used as a foundation for works to come.

October 2021

Kailong Zhang
Qun Chen

Organization

International Committee

ICIS2021-Fall

General Chair

Yanning Zhang	Northwestern Polytechnical University, China

Conference Co-chairs

Jiangbin Zheng	School of Software, Northwestern Polytechnical University, China
Simon Xu	Algoma University, Canada

Program Co-chairs

Kailong Zhang	School of Computer, Northwestern Polytechnical University, China
China Qun Chen	School of Computer, Northwestern Polytechnical University, China

Local Arrangement Chairs

Chunyan Ma	School of Software, Northwestern Polytechnical University, China
Kun Zhang	School of Software, Northwestern Polytechnical University, China

Finance Chairs

Wei Zheng	School of Software, Northwestern Polytechnical University, China
Li Wang	School of Software, Northwestern Polytechnical University, China

Publicity Chairs

Tao Zhang	School of Software, Northwestern Polytechnical University, China
Qianru Wei	School of Software, Northwestern Polytechnical University, China
Wenqian Shang	School of Computer Science and Cybersecurity, Communication University of China, China

Registration Chairs

Yin Ming	School of Software, Northwestern Polytechnical University, China
Joyce Xiao	School of Software, Northwestern Polytechnical University, China

Program Committee Members

Jianqi An	China University of Geosciences, China
Yasmine Arafa	University of Greenwich, UK
Gilbert Babin	HEC Montréal, Canada
Antoine	BossardKanagawa University, Japan
Luigi Buglione	Engineering.IT/ETS, Italy
Alberto Cano	Virginia Commonwealth University, USA
Victor Chan	Macao Polytechnic Institute, Macao
Yoonsik Cheon	University of Texas at El Paso, USA
Marc Cheong	University of Melbourne, Australia
Dickson K. W. Chiu	The University of Hong Kong, Hong Kong
Morshed Chowdhury	Deakin University, Australia
Anthony Chung	DePaul University, USA
Xiaohui Cui	Wuhan University, China
Josh Dehlinger	Towson University, USA
Mario Doeller	FH Kufstein Tirol, Austria
Hongbin Dong	Harbin Engineering University, China
Weiwei Du	Kyoto Institute of Technology, Japan
Lydie Du	Bousquet LIG, France
Yucong Duan	Hainan University, China
Zongming Fei	University of Kentucky, USA
Katsuhide Fujita	Tokyo University of Agriculture and Technology, Japan
Koji Fujita	Cyber University, Japan
Naoki Fukuta	Shizuoka University, Japan
Miguel A.Garcia-Ruiz	Algoma University, Canada
Cigdem Gencel	Free University of Bozen-Bolzano, Italy
Jiayu Gong	Shanghai Development Center of Computer Software Technology, China

Takaaki Goto	Toyo University, Japan
Jian Guan	University of Louisville, USA
Rafik Hadfi	Kyoto University, Japan
Robert Hammell	Towson University, USA
Hidehiko Hayashi	Hokusei Gakuen University, Japan
Wen-Chen Hu	University of North Dakota, USA
Naohiro Ishii	AIIT, Japan
Yuji Iwahori	Chubu University, Japan
Motoi Iwashita	Chiba Institute of Technology, Japan
Kazunori Iwata	Aichi University, Japan
Juyeon Jo	University of Nevada, Las Vegas, USA
Pankaj Kamthan	Concordia University, Canada
Keiichi Kaneko	Tokyo University of Agriculture and Technology, Japan
Sungwon Kang	Korea Advanced Institute of Science and Technology, South Korea
Mohamad Kassab	The Pennsylvania State University, USA
Masashi Kawaguchi	Suzuka College, Japan
Adel Khelifi	Abu Dhabi University, United Arab Emirates
Donghyun Kim	Georgia State University, USA
Jong-Bae Kim	Soongsil University, South Korea
Yanggon Kim	Towson University, USA
Hidetsugu Kohzaki	Kyoto University, Japan
Cyril S. Ku	William Paterson University, USA
Jay Ligatti	University of South Florida, USA
Weiguo Lin	Communication University of China, China
Chuan-Ming Liu	National Taipei University of Technology, Taiwan
Man Fung Lo	The Education University of Hong Kong, Hong Kong
Chaoying Ma	University of Greenwich, UK
Jixin Ma	University of Greenwich, UK
Mohamed Arezki	M'Hamed Bougara University, Algeria
Huaikou Miao	Shanghai University, China
Jose M. Molina	Universidad Carlos III de Madrid, Spain
Akito Monden	Okayama University, Japan
Ahmed Moustafa	Nagoya Institute of Technology, Australia
Tetsuya Nakatoh	Nakamura University, Japan
Ben C. K. Ngan	Worcester Polytechnic Institute, USA
Hiroki Nomiya	Kyoto Institute of Technology, Japan
Kazuya Odagiri	Sugiyama Jogakuen University, Japan
Toshiaki Omori	Kobe University, Japan
Takanobu Otsuka	Nagoya Institute of Technology, Japan
Athanasios Paraskelidis	University of Portsmouth, UK
Chang-Shyh Peng	California Lutheran University, USA

Contents

List of Contributors

Wissem Abbes University of Sousse, ISITCom, Sousse, Tunisia; REGIM-Lab.: REsearch Groups on Intelligent Machines, University of Sfax, National Engineering School of Sfax (ENIS), Sfax, Tunisia

Adel M. Alimi REGIM-Lab.: REsearch Groups on Intelligent Machines, University of Sfax, National Engineering School of Sfax (ENIS), Sfax, Tunisia

Kyung Mi Bae Department of International Business and Trade, School of Global Convergence Studies, Inha University, Incheon, Korea

Changhwa Baek Department of Industrial Management Engineering, Daejin University, Pocheon, Gyeongi-do, Korea

Sujeong Baek Department of Industrial and Management Engineering, Hanbat National University, Daejeon, Republic of Korea

Su In Chea GL Company, Busan, Republic of Korea

Ji Woo Choi Seogyeong University, Seoul, South Korea

Ka Cheng Choi School of Applied Sciences, Macao Polytechnic Institute, Macau S.A.R., China

Ioannis Daramouskas Computer Technology Institute and Press "Diophantus", Patras, Greece

Amir Hussain Edinburgh Napier University, School of Computing, Edinburgh, UK

Sio Kei Im Macao Polytechnic Institute, Macau S.A.R., China

Ruize Jia School of Applied Sciences, Macao Polytechnic Institute, Macau S.A.R., China

Vaggelis Kapoulas Computer Technology Institute and Press "Diophantus", Patras, Greece

Zied Kechaou REGIM-Lab.: REsearch Groups on Intelligent Machines, University of Sfax, National Engineering School of Sfax (ENIS), Sfax, Tunisia

DaeSoo Kim Korea University Business School, Seoul, Korea

Dong Oh Kim Department of Industrial and Management Engineering, Hanbat National University, Daejeon, Republic of Korea

Seo Young Kim College of Business Administration, Inha University, Incheon, Korea

Seungbeom Kim College of Business Administration, Hongik University, Seoul, Korea

Tae Hee Kim Airport Passenger Service Incheon, Korean Air, Incheon, Korea

Youn Sung Kim College of Business Administration, Inha University, Incheon, Korea

Chan Tong Lam School of Applied Sciences, Macao Polytechnic Institute, Macau S.A.R., China

Hoo-Ki Lee Department of Cyber Security Engineering, Konyang University, Nonsan, Korea

Jin-Hee Lee Department of Business Administration, Soongsil Cyber University, Seoul, Korea

Jung Jae Lee Department of Entertainment and Art Management, Soongsil Cyber University, Seoul, Korea

Seo Jin Lee Department of Industrial and Management Engineering, Hanbat National University, Daejeon, Republic of Korea

Yoo Jeong Lee Seogyeong University, Seoul, South Korea

Dimitrios Mitroulias Computer Technology Institute and Press "Diophantus", Patras, Greece

Hyun Woo Nam Seogyeong University, Seoul, South Korea

Koon Kei Ng School of Applied Sciences, Macao Polytechnic Institute, Macau S.A.R., China

Keontaek Oh Korea University Business School, Seoul, Korea

Michael Paraskevas Computer Technology Institute and Press "Diophantus", Patras, Greece

Min Seo Park College of Business Administration, Inha University, Incheon, Korea

Wonhyung Park Department of Information Security Engineering, Sangmyung University, Cheonan, Korea

Isidoros Perikos Computer Technology Institute and Press "Diophantus", Patras, Greece

Sae Young Shin Seogyeong University, Seoul, South Korea

Eunil Son Gyeongnam 6th Industrialization Support Center, Gyeongnam, Korea

Yapeng Wang School of Applied Sciences, Macao Polytechnic Institute, Macau S.A.R., China

Xu Yang School of Applied Sciences, Macao Polytechnic Institute, Macau S.A.R., China

Na Hyeon Yu Department of Industrial and Management Engineering, Hanbat National University, Daejeon, Republic of Korea

Automatic Speech Recognition for Portuguese with Small Data Set

Yapeng Wang[1(✉)], Ruize Jia[1], Chan Tong Lam[1], Ka Cheng Choi[1], Koon Kei Ng[1], Xu Yang[1], and Sio Kei Im[2]

[1] School of Applied Sciences, Macao Polytechnic Institute, Macau S.A.R., China
{yapengwang,P1708990,ctlam,rebeccachoi,bng,xuyang}@ipm.edu.mo
[2] Macao Polytechnic Institute, Macau S.A.R., China
marcusim@ipm.edu.mo

Abstract. Voice recognition has become more and more popular in various systems and applications. To further promote Macau tourism worldwide, a mobile Macau tourism APP is being developing that supports voice control to facilitate Portuguese users. Consequently, this paper is about the research and implementation of an Automatic Speech Recognition (ASR) engine for Portuguese language. In this research, three well-known open-source ASR platforms were evaluated and compared. The complete ASR development procedure using Kaldi platform is discussed. Due to the limitation of collected voice data, a novel few-shot learning and transfer learning is implemented in this project. The final model achieved a stable 95.25% accuracy which is good enough for production use. The novel technics implemented in this research can be used for ASR trainings with limited training data and can be extended to a wide range of applications in the future.

Keywords: ASR · Portuguese voice recognition · Few-shot learning · Transfer learning

1 Introduction

With the boom of artificial intelligence technology, numerous research directions like speech recognition, image recognition, and data mining, are going through a period of rapid and revolutionary development. As a result, in the area of speech recognition, technology giants like Google, Apple and Microsoft had already developed robust speech recognition engines for billions of daily active users.

However, issues related to privacy, security, and even politics led by technological barriers will undoubtedly turn into a significant disadvantage for the stability of local speech recognition services. Admittedly, these tech giants are providing some API for custom software development, but the core technology is still confidential and cannot be accessed locally. Not only the service providers could cause these potential problems, but also local users are raising high demand for speech recognition functions. That is why it is important to achieve a localized speech recognition engine.

Currently, there are several platforms for speech recognition development, such as TensorFlow [1], Kaldi [2] and HTK [3]. For this project, comparison needs to be made

R. Lee (Ed.): ICIS 2021, SCI 1003, pp. 1–13, 2022.
https://doi.org/10.1007/978-3-030-90528-6_1

within these development tools. Explicitly, for test purpose, all tools will produce a model using the same dataset. The most efficient and convenient platform will be the most suitable to work on.

Furthermore, similar project of other languages can also become references to improve the outcome of the project. Such as English Speech Recognition (LibriSpeech) [4], Brazilian Portuguese Speech Recognition [5] and Chinese Speech Recognition (THCHS30) [6]. The interesting thing is, although Portuguese is different from English and many other languages in various aspects, it is still possible to unify the pronunciation in the computer form as we already have the IPA (International Phonetic Alphabet) as a standard. ARPABET [7] was designed for Germanic languages such as English but can still be used in other languages. On the other side, SAMPA [8] was designed for multiple language using the same interpretation. Both transcriptions have the native support from Kaldi, and they are considered as the desired approaches. In addition, in order to achieve quick access with different functions in the Macau Tourism app, keyword search will be another powerful tool to help identify the user's desire.

There are several key techniques used in this research, including:

- Mel-frequency cepstral coefficients (MFCCs)

  Mel-frequency cepstral coefficients (MFCCs) are coefficients that collectively make up a Mel-frequency Cepstrum (MFC). MFC is a representation of the short-term power spectrum of a sound, based on a linear cosine transform of a log power spectrum on a nonlinear Mel scale of frequency [9]. The meaning of MFCCs is basically observing the sound at the scale of the logarithm, which degrades the sound and makes its feature more significant. MFCCs are usually calculated during the feature extraction stage of Automatic Speech Recognition (ASR) training.
- Time delay neural network (TDNN)

  Time delay neural network (TDNN) is a multilayer artificial neural network architecture whose purpose is to 1) classify patterns with shift-invariance, and 2) model context at each layer of the network [10].

  TDNN is implemented in Kaldi as it provides a good effect dealing with poor audio quality or under noisy environment. Because the TDNN recognizes audio at the phonemes level and independent of position in time, it has superior performance over the static classification [11].
- Data Augmentation

  As one of the approaches in few-shot learning, data augmentation increases the size of the dataset by changing a few characteristics from the original data. It is used to alleviate readily overfitting during the machine learning process [12]. In the field of speech recognition, data augmentation techniques like noise injection, shifting time, changing pitch, and changing speed are used to synthesis audios.
- Transfer Learning (TL)

  Transfer learning (TL) is a research problem in machine learning (ML) that focuses on storing knowledge gained while solving one problem and applying it to a different but related problem [13]. It holds the "Copy-Paste" mechanism. Numerous works were achieved by applying parts of the refined neural network to a project with a similar purpose. Transfer learning not only can be time-saving but also realize a decent outcome. As a few-shot learning approach, transfer learning can enhance performance when faced with a small dataset.

Algorithms are available for transfer learning in Markov logic networks and Bayesian networks as applications of Transfer Learning [14, 15].

No matter what development tool to implement, a speech corpus is required to gather all the training material to build an ASR system. The procedure for constructing a speech corpus is shown in Fig. 1. It consists of four sections. In the audio collection section, multiple sources are suggested to obtain audio data, such as recordings of students, web scraping, or broadcast recordings. After the stage of audio collection, a normalization process is required, which is audio pre-processing. All the audio files should be converted to mono, 16 bit, 16 kHz, following a consistent naming sequence. Other processing involving amplification, noise reduction is optional depending on the quality of the source audios. Entering the next stage, transcribing (also as known as labeling) should be done to determine the text corresponding to the collected audio. Crowdsourcing is usually considered a viable approach, especially when a large amount of audio data is provided to process. Currently a mobile App has been developed to let student collect voice samples and transcribing tasks. The last step is to combine the standardized audios and their text representations while following a specific format to build the speech corpus.

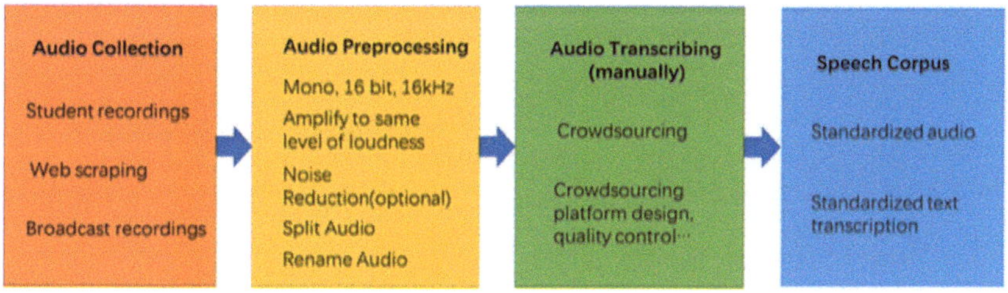

Fig. 1. Procedure of constructing a speech corpus

After constructing the speech corpus, the ASR development tool would be able to extract the information required form the corpus and train a model out of it. Depending on the ASR development tool, different operations and algorithms may be required to be implemented during model training period.

Finally, keyword searching should be implemented by create indices on the command word occurrences to respond to certain user commands.

2 System Design

This system works as a Portuguese speech recognition engine, interpreting Speech into text transcriptions. Firstly, it accepts the Portuguese speech inputs, analyzes the spectrum through a neural network. (In specific, this process is called TDNN online decoding.) After online decoding, the system would generate the transcriptions within the corpus holding the highest possibility. Finally, the output data generated by the system could be varied depending on the receiver, which relates to the system architecture and will be illustrated in the following sections.

This system operates in a web server, providing a data interface to receive inputs and return outputs. In one of the situations, a user speaks to the microphone on his/her phone, sending the Portuguese speech to a server. Explicitly, this server act as a median to request speech recognition from the speech recognition engine over TCP to a web API built in the engine. After the engine generates the Portuguese transcriptions, it returns the result through the web API. Subsequently, the median server analysis the result and send back some instructions to the APP. Another situation could be the engine is localized in the APP, receiving inputs directly from the microphone and return the result to the client.

The procedures of System Design and Development Logic are well defined as shown in Fig. 2, the detailed process will be explained in the following chapters.

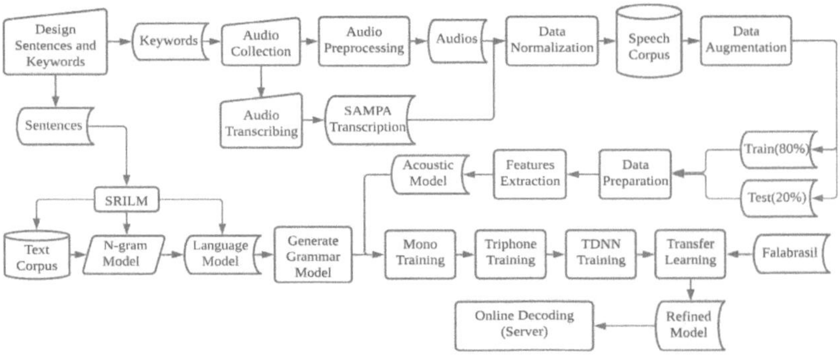

Fig. 2. System design and development logic

Generally, to develop a well-refined Portuguese ASR model, an acoustic model and a language model are required. Start from the easiest part, to construct a Language model, a rich-texted corpus is needed to get a statistical view of the recognition scope. Furthermore, to develop an acoustic model, a speech corpus is needed to extract the physical acoustic features of the recognition elements. To improve the effectiveness of these models, data normalization and augmentation should be performed. By combining these two models, an initial model is formed to be refined by various approaches like mono/triphone/TDNN training, and transfer learning. At the end of the day, a refined model (which is likely to hold an accuracy above 90%) should be deployed to a server to realize online decoding.

3 Platform Selection

There are three popular ASR development platforms: TensorFlow, Kaldi, and HTK Toolkit. Our first task is to find the most suitable tool to cooperate with and develop a powerful ASR system. The existing English dataset "Speech Command Dataset" has been utilized for comparison between TensorFlow, Kaldi, and HTK.

- Google's Speech Command Dataset
  Comparisons have done between Kaldi, TensorFlow, and HTK by training the same dataset, Google Speech Command Dataset [16]. The dataset contains words like "backward", "bed", "bird", "cat", "dog", "down", "eight", "five", etc. (about 100,000 occurrences in total). It intends to provide a corpus for the "Simple Audio Recognition" project in TensorFlow. As a result, TensorFlow can effortlessly execute it without manual modification. But for other development environments, this dataset needs to fit in the format accordingly to get implemented. One benefit of using this dataset is that not much effort is required in the normalization process as audios following the same data format. Moreover, the training target is achievable as it solely consists of simple individual words.
  Portuguese corpus is not involved in the comparison because: a) In the perspective of benchmarking, the effect of using training material from different languages is merely the same, the differences only appear in the training material. b) As a minority language, European Portuguese has few corpora to implement. On the one hand, the training material won't be enough if building a custom corpus from scratch. On the other hand, English training materials are more mature and easier to be implemented.
- Comparison between Kaldi, TensorFlow and HTK
  All works are done under the Linux operating system within an adequate barebone system. To avoid environmental issues and potential overlap between these tools, Anaconda, as a predominant program, simplify the deployment process by creating discrete virtual environments, has been implemented.

The Speech Command Dataset is used to test all three programs for its usability and reliability. TensorFlow has up-to-date tutorials based on Speech Recognition and provides fantastic visualization effects. However, the DNN algorithm in TensorFlow is used to classify words and may not be reliable with a corpus contains too many categories.

The idea to develop a DNN just using the raw audio from scratch is also not realistic. First, numerous data is required if work without feature extraction stage, this is achieved by companies that have such technical and financial strength. Second, the target of this project's desired system is focus on speech command and DNNs using raw audio usually serve the purpose of speech recognition in the scope of the whole language. Nevertheless, DNN can still be used as a part of the ASR development to further increase robustness.

Kaldi is a fully developed ASR system inherited the HMM algorithm from HTK and became popular in recent years. But the downside of Kaldi is that it is hard to manipulate. Explicitly, it doesn't have any visualized interface even in the terminal, which totally departs from TensorFlow's style.

As for HTK, it only uses HMM and it is out of date. As a matter of fact, HTK's newest version has stuck at version 3.4.1 since 13 March 2009 [17]. Due to the complexity of the operation and the abundance of maintenance issues, the HTK toolkit is not considered a viable approach.

After all the discussion above, HTK is obviously not the choice. Hence, it's reasonable to only investigate the performance between Kaldi and TensorFlow, the comparison is as shown in Table 1.

It is not hard to discover that Kaldi holds a better training speed and accuracy, also supports HMM, TDNN even DNN in the later development. However, it can only work

Table 1. Benchmarks between development platforms

	Kaldi	TensorFlow	HTK (obsolete)
Algorithm	HMM, TDNN	DNN	HMM
Training Time	27 min	1.5h	NaN
WER	8%	12%	NaN
Portability	Linux	Windows, Linux, Mac OS	Windows, Linux, Mac OS
Robustness	Quiet environment, require DNN	Quiet or noisy environment	Quiet environment
Manipulation	Perl, C++, Python, bash, etc.	Python, Java, C++, etc.	bash
Latest ver.	2021	2021	2016

under the Linux operating system, and its performance under a noisy environment still demands further customization. Moreover, it uses multiple programming languages to form its entire ASR system.

On the contrary, TensorFlow is more versatile and support almost all operating system, natively provides the robustness under noisy environment as a benefit of its DNN algorithm.

There is a trade-off between using Kaldi and TensorFlow. Although Kaldi may not be easy for manipulation and transferring between different systems, it sustains a great accuracy and can be improved with further development and customization. One of the goals of this research is to reach a high accuracy for speech recognition, in this aspect Kaldi is more beneficial comparing to TensorFlow. As a result, Kaldi is selected as the development tool for this research.

4 Portuguese ASR Implementation

As Kaldi has been chosen as the development platform of this work, training Portuguese in Kaldi becomes the next goal. To achieve the conversion from English to Portuguese, works associated with obtaining a speech corpus need to be done. Using an existing corpus is possible, but it is not realistic spending an exorbitant price and tremendous time effort to get a corpus still required to make adaptive changes in Macau tourism. Consequently, constructing a customized corpus for Macau tourism is more appropriate.

- Keyword/Sentences Selection

  Based on the requirement of the Macau tourism APP, keywords and sentences were established from various categories including command, functions, numbers, casinos, shopping marts, hotels, attractions, and restaurants. All the keywords/sentences are designed to fit in the daily dialogues of the Portuguese. The list contains the corresponding representations in Portuguese, English, and Chinese. We map the Portuguese words to English for the readability and usability of coding.

The independent keywords and those selected from sentences would be used for the construction of speech corpus and acoustic model. On the other hand, the sentences will be used to train a language model after augmentation. A gold principle of sentence design is to produce variations as many as possible for the same meaning/purpose.

However, some difficulties are encountered such as the Portuguese words could be varied because of conjugations and other Portuguese grammars. Specifically, in Portuguese, words have two genders (génèro). Words end with 'a' are usually female (feminino), words end with 'o' or consonants are usually male (masculino), some words are both male and female (substantivo sobrecomum), but there are many exceptions. Most importantly, there are words with both forms, varies based on the context.

So, there is some uncertainty about the form of words in different situations. The solution to this problem is recording all possible forms of the words and map the meaning into English. After adopting this strategy, no decision needs to make to determine which form belongs to which situation. Nevertheless, this could potentially cause some problems when dealing with polysemy in English. Although manually choosing different synonyms may fix certain occasions, using Natural Language Processing (NLP) would solve this problem entirely with a statistical model. In fact, this is one of the limitations of the conventional approach to Speech Recognition.

Moreover, because the designed keywords/sentences is using in a conversation scenario, the subject could be varied. Thus, sentences constructed with imperative (Imperativo) mood is one of the participants in the collections as a convenient choice.

The pronunciations of words vary depending on the sentences they lie in. To generate a natural intonation such as real-life conversation, volunteer students have recorded some sentences instead of discrete words only. After that, these words were pulled away from sentences using audio editing software.

Another interesting observation is that though some Portuguese words may vary, there are parts that stay unchanged. This reduced the workload as only the changing parts requiring consideration. Also, these unchanged parts will be treated as anchors to identify the semantics by associating them with corresponding APP functions.

- Audio Collection, Preprocessing and Transcribing

  Students were volunteered to record the sentences and keywords. In summary, Hi-fidelity audios have been recorded by six native speakers using recorders. Also, 15 recordings with smartphones have been collected to fit in a practical scenario.

The native speakers and non-native speakers have distinct intonations. If the data size is efficient, it would be necessary to separate them into two types and characterize them using DNN. Especially recording a word only has an English representation, it is more appropriate to collect audios with a Portuguese accent (intonation) by a native speaker.

The recording environment isn't necessary to be completely quiet but cannot be too noisy. The recording can be split into multiple parts later, and one speaker can record all the content into one audio.

FFmpeg [18] has been used to convert all audios into the same format (16 bit 16 kHz mono wav).

For English, it's easy for the transcribe process, as there's already auto-transcribe tools developed for English. These tools can generate the ARPABET transcriptions base on the English words, and no tools existing are designed for Portuguese. However, ARPABET transcriptions can still be generated for Portuguese. In specific, the articulatory configurations [19] of the human vocal tract are fixed, which causes the sound that humans can produce is fixed. Therefore, the phonemes humans can pronounce are the same.

At the phoneme level, most languages can be identified as phonemes with intonations. ARPABET is exactly a standard of transcription built at the phoneme level, that's why multi-language ASR can be achieved by using it.

The method to generate ARPABET from Portuguese is dividing each Portuguese word into phoneme by searching the IPA representation, then based on the IPA-to-ARPABET table to convert these IPA representations to ARPABET transcriptions. Although that table does not contain all the IPA, as an auxiliary way, SAMPA-to-ARPABET can help to map Portuguese to ARPABET when IPA can't. But there are situations that no table entry is matched to the ARPABET, that's when help from Portuguese teachers or Linguists is needed to judge which pronunciations it is supposed to be. In other ways, SAMPA along can also be used as the transcriptions directly in Kaldi.

Finally, a dictionary specifies the IPA-to-ARPABET was constructed with the help of combining the two tables, after several times of testing.

To make transfer learning more effective, a Brazilian Portuguese model "FalaBrazil" [20] was used to enhance the model. But due to the requirement of data consistency, the transcription cannot use ARPABET. In fact, it requires the SAMPA transcription, which is the transcription FalaBrazil is using. One benefit of using SAMPA is that it has better support for multilingual speech recognition, thus a combination of Portuguese and other languages can be recognized at the same time.

- Data Preparation and Training

  As both acoustic and language data need to be prepared for later training process, we need to split the data into training dataset and testing dataset. The training set is 80% of the dataset, the test set is 20%. A script is developed to split the training audios and test audios from the speech corpus. After separating the training set and the test set, the acoustic data can be generated. After that, a series of operations including features extraction, making the grammar model, mono training, triphone training, decoding and alignment was executed. Finally the TDNN was used to train the data. Because the amount of training data is not enough, overfitting frequently happens, and it's not easy to manipulate the learning rate of the epochs during the training.

During the triphone training stage, the accuracy becomes stable at around 90%, but the TDNN training does not offer any improvements with the accuracy. WER (word error rate) is 9.43% with one insertion error and four substitution errors.

- Few-shot Learning

  Few-shot learning refers to the practice of feeding a learning model with a very small amount of training data, contrary to the normal practice of using a large amount of data. With an insufficient amount of training materials, few-shot learning approaches need to be adopted to improve accuracy. Data Augmentation and Transfer Learning are few-shot learning approaches suitable for the speech recognition scenario.

To further improve the accuracy, data augmentation is used to generate more data for training. For the recording audios, the conventional approaches include Noise Injection, Shifting Time, Changing Pitch, and Changing Speed, which can be effortlessly achieved by using a few lines of python code that can be found directly online. This is used to remedy the situation with small amount of dataset to avoid quick overfitting. The augmented data cannot exceed too far from the amount of original data, as experts suggested [21]. Another data augmentation approach is Google's SpecAugment technology [22]. This algorithm augment audio data in the spectrogram level, transform this problem into an image augmentation one. SpecAugment already has an implementation in Kaldi: mini-librispeech. By using these two data augmentation measures, the dataset was enlarged. Specifically, the total number of occurrences generated is 4 times of the its original size: $22 * 21 * 4 = 1684$ entries.

For the sentences, after changing the sequence of the constituent, the quantifier, and synonym representations, more entries are generated. The augment data contains more possible sentences which is 6.3 times of its original size: $126 * 6.309 = 795$ entries. By doing this, new sentences generated can be used to build a new language model.

- Transfer Learning

  Transfer learning is the improvement of learning in a new task through the transfer of knowledge from a related task that has already been learned. Generally, Transfer learning is used to migrate features from an ideal model and use a subset of its parameters to initialize a new model for a new task. It's a good way to reduce the cost and simplify the training process to train a new model base on existing models.

As Fig. 3 shows, a subset of model parameters trained on a language using the Latin alphabet is used to create a new model for a new language, written with the Cyrillic alphabet. Blue model parameters have been copied from the source model, and the green model parameters have been re-initialized from scratch [23].

Practically speaking, the well-trained model of LibriSpeech is easy to obtain and open to the public. To make it works, the transcriptions were in the format of ARPABET at first. LibirSpeech was first implemented as the model to learn from, but the training results were not as appealing as they should be. Specifically, the error rate could reach above 50%.

That's why the FalaBrazil model was involved. It's a Brazilian Portuguese ASR engine, having a greater similarity with the model. But the transcription designed in FalaBrazil is SAMPA comparing to the English-based LibirSpeech. To keep the data consistent, the transcribing needs to redo again in the SAMPA format.

The script in the RM corpus can be implemented in the transfer learning process. It was first designed to achieve transfer learning from WSJ to RM, and it needs a few files from the input model and then mixes up the data at a fitting ratio.

Finally, the model needs fine-tuning at the DNN training stage with the learning rate and epochs.

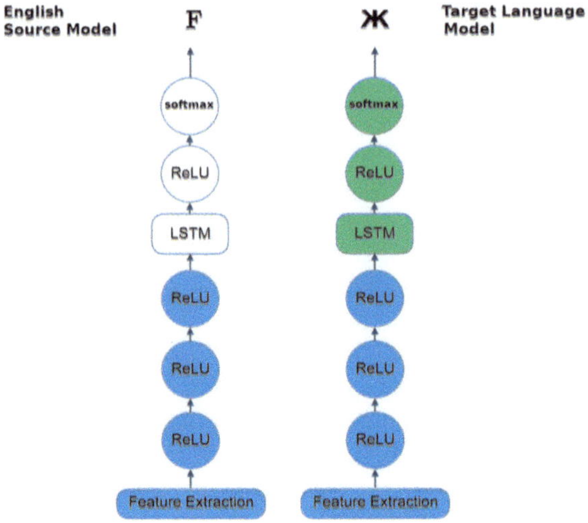

Fig. 3. The "Copy-Paste" mechanism of transfer learning

- Online interface for ASR

  One of the objects regarding data interfaces for deployment for online/offline ASR can be realized by carrying out Kaldi's built in component, online decoders.

  There are different decoders handling different inputs/outputs regarding different scenarios. By the help of these pre-built functions, no extra work is needed for developing an independent data interface. Fortunately, these decoders are capable of both deployment and decoding. For online decoding using TDNN, online2-tcp-nnet3-decode-faster was implemented, which receive WAV input on the server side. For offline decoding, online2-wav-nnet3-latgen-faster was implemented to receive input from the local microphone.

5 Results and Discussions

As mentioned previously, increasing the amount of data is the direct approach to increase accuracy. By contrast, using Transfer Training becomes the indirect approach to improve the robustness. And the process of Transfer Learning is very similar to DNN/TDNN training by omitting certain processes. In this way, fine-tuning should be comprehended to yield a positive outcome on the model. By visualize the training process and investigating the suitable level of learning rate and times of epochs, it's feasible to conclude a mathematical induction out of observations and experiments to prevent underfitting or overfitting.

Because the dimensions of the two models are different, they were combined proportionally at about 4:1 to achieve a better effect.

The WERs after all the result after implementing few-shot approaches, the improvement is significant. The last output indicates the accuracy of using online decoder. The final WER(word error rate) becomes stable at around 4.75% with no insertion error, 26 deletion errors and 54 substitution errors.

As shown in Table 2, after all the refinements, this model reached a WER of 4.75%, indicating the accuracy of 95.25%. Also, improvements have been made comparing to the former result without implementing few-shot learning. (9.43% – 4.75% = 4.68%) Overall, the robustness of the model is tested quite well.

Table 2. Voice recognition results

Stages	WER before Few-shot learning	WER after Few-shot learning	Improvement
Mono	16.98%	12.35%	4.63%
Triphone	9.43%	7.54%	1.89%
TDNN	N/A	4.75%	N/A

At this stage, the comparison of three platforms, data collection, data handling, model training, few-shot learning, model refinement, model deployment, and decoding has been performed. Explicitly, Kaldi was chosen as its superior functions, command keywords/sentences were designed and recorded to construct a speech corpus, scripts were developed to generate data in the stage of data preparation automatically, an initial acoustic model was trained based on feature extraction but got updated after few-shot learning and DNN training, the fine-tuning was performed to ensure a 4:1 ratio of transfer learning, the data interface has native support form Kaldi, thus it's easy to achieve online/local deployment.

There're certain novelties in this research, including the unique design of words/sentences when the dataset falls into a small category, interpreting IPA to ARPA-BET transcription for multilingual transfer learning, the adoption of transfer learning to a European Portuguese model etc. All of them are original ideas. Some of them may turn out to have little or no effects, but certain measures can be proved to improve the efficacy of work and recognition robustness significantly. It's worth noting that no existing similar work is done in this category for Portuguese.

Although the objects are met, some potential improvements can still be made in the future. One is to use the developed voice collection app to collect more training audios. The other thing is that a simple semantic recognition using some tactics (Anchors/Keyword Search/NLP) could further improve the recognition effectiveness.

The final refined mode is now being deploying to the server to achieve online ASR, which offers a robust European Portuguese ASR engine for Macao Tourism APP. Consequently, this could facilitate an intelligent system, further promoting Macao worldwide in the field of tourism.

6 Conclusion

In conclusion, this research initially compared various platforms for speech recognition development, Kaldi was chosen for its performance and scalability after scrutinizing operations with all three platforms. Despite the difficulties encountered in the Portuguese word selection and audio transcribing, statistical ideas were advanced to solve grammar problems. From speech corpus constructing to features extraction, from data augmentation to new LM/AM, from monophone training to triphone training, until TDNN to Transfer Learning, the model was refined and its robustness has been improved, reaching an accuracy of 95.25% with proof of online decoding test.

There is no existing work similar to this work in the field of European Portuguese ASR. This work could become a work of reference for its novelty for other languages' ASR developing. Additionally, after some improvement and modifications, the model may have a wide range of applications in the future, serving as voice assistants, audio guides, and even accessibility providers for the disabled in Macao.

Acknowledgments. This research is funded by The Science and Technology Development Fund, Macau SAR (File no. 0082/2018/A2).

Reference:s

1. Abadi, M., et al.: TensorFlow: A System for Large-Scale Machine Learning (2016)
2. Povey, D., et al.: The Kaldi speech recognition toolkit. In: IEEE Workshop on Automatic Speech Recognition and Understanding, pp. 1–4 (2011)
3. Young, S., et al.: The HTK Book (1995)
4. Panayotov, V., Chen, G., Povey, D., Khudanpur, S.: Librispeech: an ASR corpus based on public domain audio books. In: ICASSP, IEEE International Conference on Acoustics, Speech and Signal Processing (ICASSP), August 2015, pp. 5206–5210 (2015). https://doi.org/10.1109/ICASSP.2015.7178964
5. Silva, P., Neto, N., Klautau, A., Adami, A., Trancoso, I.: Speech Recognition for Brazilian Portuguese using the Spoltech and OGI-22 Corpora
6. Wang, D., Zhang, X.: THCHS-30: A Free Chinese Speech Corpus (2015)
7. Bell, A., Jurafsky, D., Fosler-Lussier, E., Girand, C., Gregory, M., Gildea, D.: Effects of disfluencies, predictability, and utterance position on word form variation in English conversation. J. Acoust. Soc. Am. **113**, 1001 (2003). https://doi.org/10.1121/1.1534836
8. Wells, J.C.: Computer-coding the IPA: a proposed extension of SAMPA
9. Muda, L., Begam, M., Elamvazuthi, I.: Voice Recognition Algorithms using Mel Frequency Cepstral Coefficient (MFCC) and Dynamic Time Warping (DTW) Techniques. 2 (2010)
10. Peddinti, V., Povey, D., Khudanpur, S.: A time delay neural network architecture for efficient modeling of long temporal contexts. In: Sixteenth Annual Conference of the International Speech Communication Association (2015)
11. Waibel, A., Hanazawa, T., Hinton, G., Shikano, K., Lang, K.J.: Phoneme recognition using time-delay neural networks. IEEE Trans. Acoust. **37**, 328–339 (1989). https://doi.org/10.1109/29.21701
12. Schlüter, J., Grill, T.: Exploring data augmentation for improved singing voice detection with neural networks. In: ISMIR, pp. 121–126 (2015)
13. Torrey, L., Shavlik, J.: Transfer learning. In: Handbook of Research on Machine Learning Applications and Trends: Algorithms, Methods, and Techniques, pp. 242–264. IGI Global (2010)
14. Mihalkova, L., Huynh, T., Mooney, R.J.: Mapping and revising Markov logic networks for transfer learning. In: AAAI, pp. 608–614 (2007)
15. Niculescu-Mizil, A., Caruana, R.: Inductive transfer for Bayesian network structure learning. In: Artificial Intelligence and Statistics, pp. 339–346. PMLR (2007)

16. Google AI Blog: Launching the Speech Commands Dataset. https://ai.googleblog.com/2017/08/launching-speech-commands-dataset.html. Accessed 13 July 2021
17. HTK Speech Recognition Toolkit. https://htk.eng.cam.ac.uk/docs/history.shtml. Accessed 13 July 2021
18. Lei, X., Jiang, X., Wang, C.: Design and implementation of a real-time video stream analysis system based on FFMPEG. In: 2013 Fourth World Congress on Software Engineering, pp. 212–216. IEEE (2013)
19. Gick, B., Wilson, I., Derrick, D.: Articulatory Phonetics. Wiley, Malden (2012)
20. Neto, N., Silva, P., Klautau, A., Adami, A.: Spoltech and OGI-22 baseline systems for speech recognition in Brazilian Portuguese. In: Teixeira, A., de Lima, V.L.S., de Oliveira, L.C., Quaresma, P. (eds.) PROPOR 2008. LNCS (LNAI), vol. 5190, pp. 256–259. Springer, Heidelberg (2008). https://doi.org/10.1007/978-3-540-85980-2_33
21. Data Augmentation for Audio. Data Augmentation | by Edward Ma | Medium. https://medium.com/@makcedward/data-augmentation-for-audio-76912b01fdf6. Accessed 13 July 2021
22. Park, D.S., et al.: Specaugment: a simple data augmentation method for automatic speech recognition. arXiv Prepr arXiv:190408779 (2019)
23. Meyer, J.: Multi-task and transfer learning in low-resource speech recognition (2019)

Service Bursting Based on Binary PSO in Hybrid Cloud Environment

Wissem Abbes[1,2(✉)], Zied Kechaou[2], Amir Hussain[3], and Adel M. Alimi[2]

[1] University of Sousse, ISITCom, 4011 Sousse, Tunisia
wissem.abbes@enis.usf.tn
[2] REGIM-Lab.: REsearch Groups on Intelligent Machines, University of Sfax, National Engineering School of Sfax (ENIS), BP 1173, 3038 Sfax, Tunisia
[3] Edinburgh Napier University, School of Computing, Edinburgh, UK

Abstract. Given the competition and rapidly-evolving market associated challenges, companies need to be innovative and agile, particularly regarding the customers dedicated web applications. Nowadays, hybrid cloud stands as an attractive solution for organizations that tend to use a combination of private and public cloud implementations, following their appropriate needs to apply the available resources and speed up execution in the most profitable ways. In this regard, deploying the new applications would certainly entail placing and devoting some components to the private cloud option, while reserving others to the public cloud option. For this purpose, a Binary Particle Swarm Optimization (BPSO) based approach is proposed, useful for maintaining an effective service bursting optimization within the hybrid cloud framework. Using a real IBM based benchmark, the experimental results appear to reveal that our advanced approach scored results turn out to outperform those documented in the experiment works in terms of cost.

Keywords: Service Bursting · BPSO algorithm · Hybrid cloud

1 Introduction

Currently, companies are increasingly deploying cloud environments and running the relevant applications. Most often, the cloud application commonly opted for in most cases is the hybrid type, wherein, the relevant infrastructure involves two or more cloud models (public or private) behaving independently. Hybrid cloud bursting can be a commonsense strategy for dealing with the extending complexity of gigantic data assessment, especially for iterative applications. With the significant advancement noticeable in the cloud registering area, the industry is transferring and bursting its applications and processes into the cloud, given the various features and advantages the cloud circumstances turn out to offer. Thanks to the business service providing components, the Cloud Platforms are increasingly apt to deploy and implement service-based applications. Once these applications are deployed within a hybrid cloud framework, choice of the appropriate public-cloud applicable components remains an open-ended question. It is worth mentioning, however, that several parameters need be considered when deciding on the

R. Lee (Ed.): ICIS 2021, SCI 1003, pp. 14–26, 2022.
https://doi.org/10.1007/978-3-030-90528-6_2

application components to be transferred to the public cloud. Among these parameters one could well cite privacy, QoS, security, communication costs as well as the relocation associated hosting costs. The bursting related problem, subject of the present paper, lies in the NP-Hard type of problem [1], primarily due to the inexistence of optimal algorithms enabling to solve them in a prompt way. With regard to the present work, each service is deployed as part of either private cloud or public cloud, which justifies our opting for the BPSO. Accordingly, each particle's respective position would be equal either to 0 (if the service is liable to deployment in the private cloud), or to 1 (if the service proves to be liable to deployment in the public cloud). In this context, a BPSO based approach is advanced, whereby, an approximate result of the optimal solution can be provided.

The remainder of the present conference paper involves the following sections. Section 2 introduces highlights of the major works dealing with cloud data-placement strategies, enumerating the major differences distinguishing the relating achievements, along with the present research contribution and achieved findings. Section 3 involves a representation of Service Based Application (SBA), with illustrative graphs and a formulation of the SBA placement problem within hybrid cloud contexts. As for Sect. 4, it includes a presentation of our advanced BPSO placement-problem solving approach. As regards the experimental analyses and evaluative procedures, they make subject of Sect. 5, while the ultimate section englobes the major concluding remarks and prospective venues for a potential research work.

2 Related Works

In this section, we consider examining a set of state-of-the-art works dealing with the resource placement issue in public, private, and hybrid clouds. Each work should represent a particular approach useful for selecting a particular service package, while accounting for some important relevant criteria. Thus, the present section involves a description of the major approaches treating the major Cloud Computing resource-management optimization issues. In this regard, Yusoh and Tang have conducted a series of studies [2, 6, 7] dealing with the SaaS component placement, including the problem of service components' mapping to VMs and storage systems. The proposed solutions involve a Cooperative Coevolutionary GA (CCGA) [6] that serves to split populations into groups aimed to optimize the computation and storage of allocations. Although the CCGA attained results appear to outperform those reached via classical GA, they are even further refined by RGGA and turn out to achieve even better results than those reached via the First Fit Descent (FFD) algorithms. In [4], Wada et al. suggest an optimiztion scheme, dubbed E3, in a bid to solve the SLA-aware Service Composition (SSC) problem. Accordingly, the E3 framework serves to define a service composition model and helps provide two heuristic algorithms, labelled E3 Multi-Objectives GA (E3-MOGA) and Extreme-E3 (X-E3), enabling to solve the SSC problem. The relevant experiments end up displaying rather effective results compared to the NSGA-II [8]. Similar objectives have been targeted by SanGA [5], a Self-adaptive network-aware GA, to help solve the service composition problem. Among the optimization targeted objectives are reducing latency and price. In this regard, SanGA proves to record the best performance among a set of tested algorithms, such as standard GA, NetGA [9], random and Dijkstra. A higher CC layer BPaaS problem is tackled by Li et al. in [4], who undertook to solve a service location problem related to the cloud logistics domain via PSO. Accordingly, the optimization objective is a weighted sum of time, price, availability, and reliability. Ultimately, their proposed solver turns out to be capable of locating the minimum value solution at a faster pace than the GA solver. For Kaviani et al. [10], a software service placement in hybrid cloud environments design is put forward, with the aim of boosting latency without increasing costs. In this same vein of thought, Charrada et Tata [11] proposed a special FBR (Forward Backward Refinement) algorithm useful for effectively placing service-based applications in hybrid Clouds. The objective behind setting up this algorithm consists in minimizing the costs generated by deploying the cloud related services. Similarly, Abbes et al. [12] proposed a novel hybrid cloud placement optimization approach based on genetic algorithm (GA). The optimization target strategy lies in minimizing the public cloud service deployment cost. The proposed approach achievement appears to outperform the FBR algorithm accomplished achievement [11] in terms of cost, recording a result close to optimal. On the other hand, the authors in [13] put forward a cloud bursting algorithm that accounts for time-varying electricity rates with private clouds along with the time-invariant rental rate of several public clouds. To this end, a special VMs placement metaheuristic approach was proposed in [14], for the sake of minimizing energy consumption and SLA violations within the cloud data center.

After intense biographical search, one comes to conclude that the majority of the relevant literature conducted works appear to consider several criteria (bandwidth of control and transmission time, cost, performance, etc.), which seem to apply exclusively either to the public or to the private cloud associated investment cases. Indeed, only a few works appear to opt for implementing a hybrid cloud relevant solution. In the latter's case, the cost of interdepartmental communication turns out to depend highly on the service placement strategies (whether on a private or a public cloud system). Indeed, the cost of an intra-cloud based communication (i.e., within private or public cloud systems) proves to differ remarkably from that of an inter-cloud based one (regarding the hybrid option). As commonly recognized, the inter-cloud communication cost is usually higher than that of intra-cloud communication. Hence, the entirety of service placement approaches considered in this context would focus on this particular feature. With regard to our particular approach, a clear distinction is established between public based communication and private based one. In this context, different approaches [15, 16] and [10] have been considered for a specific hybrid cloud based architecture with reduced user investment to be achieved. The objective has been to minimize costs by allowing users to decide on which resources to opt for, in such a way as access to resources turns out to be transparent, while enhancing scalability, reliability and minimizing costs. It is worth noting, however, that on optimizing the resource placement costs, various approaches do not seem to consider communication flow between the different clouds' parameters, which involve significantly important costs.

In an attempt to account for such shortcomings, a BPSO based approach, relevant to service placement on a hybrid cloud, is advanced within the scope of the present paper. Our focus of interest is primarily laid on treating the service placement generated costs (hosting cost, communication between services cost).

3 Applications Placement on Hybrid Cloud

3.1 Problem Formulation

It is worth highlighting that the Service-Based Application (SBA) in hybrid cloud, subject of our previously proposed mixed framework involving both structures [12], which we consider as a composition of private and public cloud systems.

At this level, SBA is a set of basic services aimed at providing complex and flexible features special environments, widely scattered by dispersed ranges and arrays of services that each environment maintains. Yet, SBA could still be deployed via public cloud in certain cases, mainly, when: (1) the deployed applications prove to require greater resources that private cloud fails to provide on its own; (2) private cloud turns out to be unable to satisfy a new deployment request; or (3) the deployed applications release resources indicating that a new application need be implemented to release allocated resources from the public cloud based data. In this respect, we have specified a defining threshold determining the appropriate case when appeal to the public cloud is mostly required. This threshold can be quantified via hosting units, hence, the notation HQ designated to refer to hosting quantity. Accordingly, the quantity of resources required for any selected services has to be greater or equal to HQ, as indicated below:

$$Minimize : H + PC + HC \tag{1}$$

$$Subject\ to :$$

$$\sum\nolimits_{i=1}^{n} h(si) \times l(si) \geq HQ \tag{2}$$

$$Where :$$

$$H = \sum\nolimits_{i=1}^{n} \alpha \times h(si) \times l(si) \tag{3}$$

$$PC = \sum\nolimits_{e=<si,sj>\in E} \beta 2 \times c(e) \times l(si) \times l(sj) \tag{4}$$

$$HC = \sum\nolimits_{i=1}^{n} \sum\nolimits_{js.t.e=<si,sj>\in E} \beta 1 \times c(e) \times l(si) \times (1 - l(sj)) \tag{5}$$

Where: (1–6) stands for the objective function (minimize H (Hosting cost) + PC (Public Communication cost) + HC (Hybrid Communication cost));

(2): denotes a constraint equation that represents the sum of hosting quantity (HQ) of the public cloud deployed services, which have to be greater than or equal to HQ (minimum threshold). HQ is a value assigned by a resource request-signal case. The need for a public cloud based resource can be quantified in terms of amount of hosting units (units of platform resources required). Once a public cloud request is triggered, one has to decide on the application services necessary to opt for, to be deployed as part of a public cloud service. In this case, the quantity of required platform resources relevant to satisfying the selected services has to be either equal to or greater than that of HQ;

(3): is the sum of hosting service costs as deployed on public cloud;

(4): designates the sum of publically made communications (as established across the public cloud deployed services);

(5): is the sum of hybrid sustained communications (as maintained through the public cloud deployed services, along with those ensured via private cloud).

Table 1. Details of the problem formulation abbreviation and the respective denotations.

Abbreviation	Denotation
H	is a hosting function that associates a positive number to each service, representing its need in hosting quantity of resources for its deployment
C	is a communication function that associates to each edge e = < s1, s2 > a positive number representing the communication rate established between s1 and s2
L	is a location function that allocates 0 to each private cloud deployed service, and 1 if the service is deployed via public cloud
β1	is the cost of a communication unit maintained between the public and the private clouds
β2	is the cost of a communication unit established within the public cloud
H	is the sum of hosting costs of services as deployed within public cloud. Indeed, the expression $\alpha \times h(si) \times l(si)$ (where α is the cost of a resource hosting unit, h(si) is hosting quantity of service si and l(si) takes the value 1 if si pertains to the public cloud, and 0 otherwise), which is equal to the si hosting cost if this service is maintained via the public cloud
PC	denotes the sum of publically established communications (communications between services deployed within the public cloud). Indeed, there is a public communication maintained between si and sj if they are both deployed within the public cloud service, $l(si) \times l(sj)$ equal 1
HC	stands for the sum of hybrid communications (communications between services deployed within the public cloud and those deployed within the private cloud). Indeed, there is a hybrid communication between si and sj if one of them is deployed via public cloud and the other is deployed via private cloud, since, either $l(si) \times l(1-(sj))$ equals 1 or $l(sj) \times l(1-(si))$ equals 1

3.2 Study Case

The structured process relates to opening a bank account [17], as illustrated in the Business Process Model and Notation (BPMN [18]) diagram, depicted through Fig. 1, below.

The SBA representation, as illustrated on Fig. 1, can also be modelled partially in the form of a graph, as shown on Fig. 2, below, where services and gateway nodes are represented by graph nodes and inter-service connections/transitions by edges. Accordingly, nodes are identified by numbers, and characterized with amount of resource hosting

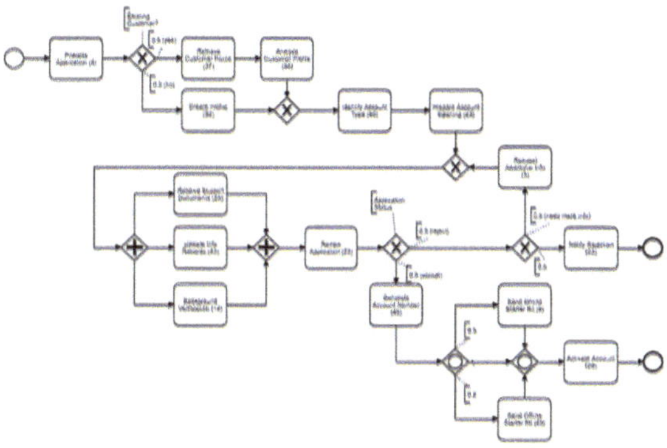

Fig. 1. An SBA application sample, as modelled after the structured process

units. Edges are characterized with some communications established, referring to the amount of traffic transferred on the considered edge.

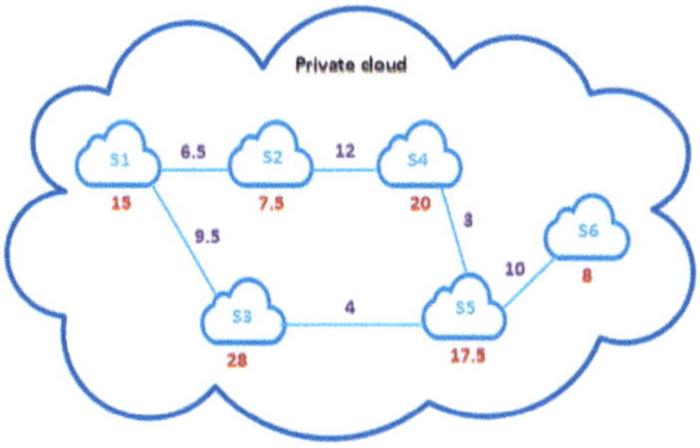

Fig. 2. SBA graph relevant to a bank account opening

The SBA graph, as modelled through Fig. 2, appears to comprise distinct services. Each service englobes a hosting quantity, and each edge involves a quantity of communication unit. SBA, as deployed in a hybrid cloud, is represented by a graph, as shown on Fig. 3, where some services are deployed via a public cloud while others are maintained via a private cloud.

The cost relevant to the model applications deployment, as illustrated through Fig. 3, for $\alpha = 40$, $\beta 1 = 10$, $\beta 2 = 5$, is determined in the following way: $40 \times (17.5 + 8) + 10 \times (3 + 4) + 5 \times 10 = 1140$. Note that, at this level, no communication and hosting cost considerations are being accounted for within the private cloud case. Indeed, assuming that a company has its proper private cloud structure, these costs would not be computed since they are generated on a private cloud basis.

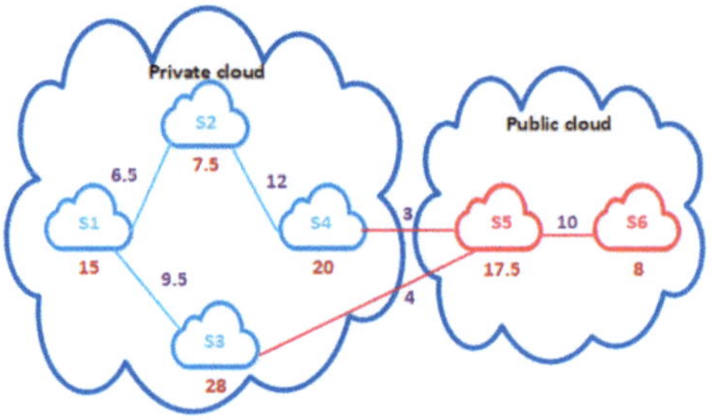

Fig. 3. Bank account opening as deployed within a hybrid cloud architecture

4 The Proposed BPSO Approach

In our context, the service placement process, from private cloud to public cloud, is designed via a binary model, wherein, each service highlights its proper location (0 if it is within the private cloud, and 1 if is maintained via public cloud). At this level, we consider implementing the PSO scheme. It is worth recalling in this context that Kennedy and Eberhart [19] were pioneers in proposing a discrete binary version of PSO. In their devised model, a special particle will decide on the "true" or "false", "yes" or " no" options etc., and these binary values may well denote special representations of real values within a binary search space. In what follows is a presentation of the BPSO relevant formula. Updating a particle's velocity is maintained via the following formula:

$$v_i^d = w * v_i^d + c_1 r_1 * (pbest_i^d - x_i^d) + c_2 r_2 * (gbest^d - x_i^d) \tag{6}$$

As to the particle's position updating, it conforms to the following formula:

$$Sig(v_i^d) = \frac{1}{1 + e^{-v_i^d}} \tag{7}$$

if $Sig(v_i^d) > r3$, then

$$x_i^d = 1. \tag{8}$$

else

$$x_i^d = 0. \tag{9}$$

Where:
W = stands for inertial weight;
$v_i{}^d$ = represents velocity for particle i at dimension d;
c_1 = designates the acceleration constant;
r_1 = is a random value;
$x_i{}^d$ = denotes position for particle i at dimension d;
pbest = is the best previous position of the ith particle;
c_2 = is an acceleration constant;
r_2 = is a random value;
gbest = refers to the best global position of all particles;
r_3 = is a random value.

5 Experimental Evaluation

5.1 IBM Dataset

For evaluative purposes, of our advanced approach, an appeal has been made to a real IBM-ased dataset [20], involving 560 BPMN. A selection of graphs, incorporating some nodes, ranging between 11 and 20, was also considered, along with 10 randomly selected SBA graphs reflecting the architecture based compositions of services. Some of the graphs are dense, while others are sparse. Density is expressed in terms of percentage computed as 100 times the ratio of the number of edges to the number of all possible edges. The considered graph relevant characteristics are depicted on Table 2, below.

Table 2. The selected graph's characteristics

Graphs	Nodes	Edges	Hosting needed	Density
G1	20	19	469	10%
G2	17	28	521	20%
G3	18	46	418	30%
G4	11	22	254	40%
G5	16	60	413	50%
G6	14	55	332	60%
G7	13	55	319	70%
G8	19	137	570	80%
G9	15	95	363	90%
G10	12	66	297	100%

Actually, the service-based applications are usually depicted in the form of graphs, which may appear either as sparse, dense or full, wherein, density represents a very important criterion. By means of illustration, an assessment instance of our devised algorithm is depicted. In G7, for instance, the possible number of edges is $(13 * 12)/2 = 78$, and we have selected 55 edges, yielding a density range of: $55/78 = 70\%$. As indicated on Table 2, a selection of graphs with different densities has been chosen, representing different compositions of service types. Based on the set benchmark designation, 10 graphs have been selected displaying different densities ranging from 10% to 100%. Regarding the problem modelling procedure, it has been implemented via CPLEX [21] (an optimization software package developed by IBM to help solve any integer programming problems likely to emanate) for the optimal solution to be effectively computed. For comparison purposes, the FBR (Forward Backward Refinement) algorithm, originally developed by [11], has been applied, as an approximate service-placement algorithm, while the GA (Genetic algorithm), as developed by [12], would stand as a GA-based placement optimization approach. Thus, we have conducted more than 2150 experiments, and the attained findings prove to demonstrate that our suggested

BPSO architecture turns out to yield very promising results, with respect not only to the sparse graphs, but also to the dense ones. Indeed, the entirety of the BPSO algorithm scored results appear to outperform noticeably both of the FBR and GA algorithms achieved ones, within the same response time interval. For illustration purposes, some of the recorded results are presented below.

Note that for all the ensuing experiments, the parameter values (as delivered by service providers) have been set as follows:

- $\alpha = 40$ is the hosting units relevant coefficient;
- $\beta 1\ 1 = 20$ is a hybrid communication associated coefficient;
- $\beta 2 = 10$ is a public communication related coefficient.

6 The Reached Results

In our simulation procedure, three among the ten Table 2 figuring graphs have been considered. They are, consecutively, a sparse graph (Fig. 4), a complete graph (Fig. 5) and a dense graph (Fig. 6), with varying HQ value ranges (ranging from 10% to 90% of the considered graph hosting quantity). In our experimental results, three HQ values have been applied (30%, 50% and 70%).

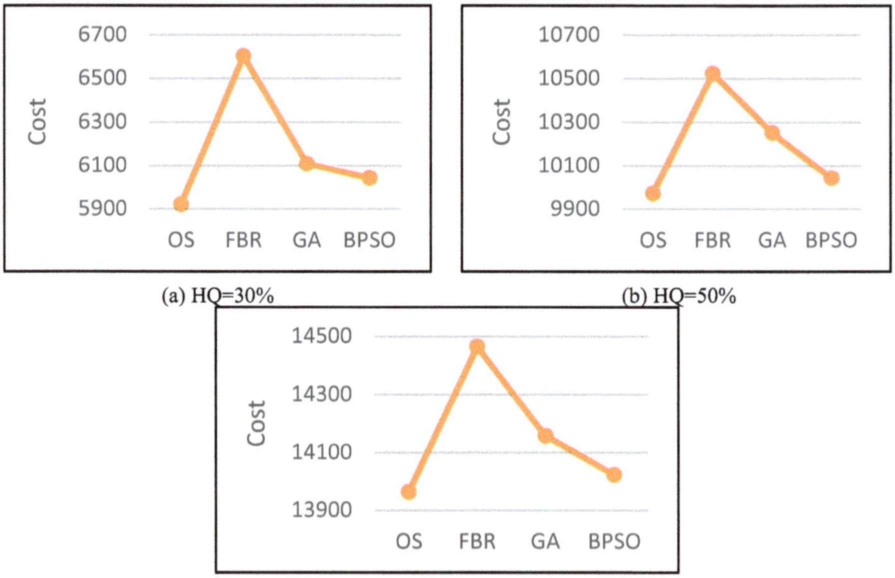

(a) HQ=30%

(b) HQ=50%

(c) HQ=70%

Fig. 4. Cost comparison involving BPSO, GA, FBR and OS (appearing on graph G1)

Figure 4 depicts costs as generated by FBR, GA, BPSO and OS relevant to the G1 graph figuring on Table 2, with density ranges of 30%, 50% and 70%. As can be noted, the BPSO scored costs turn out to be consistently lower than those reached via FBR.

The recorded cost differences, prevailing between both of the BPSO and FBR appear to decrease with increased HQ values, and vice versa. These results can be justified by the fact that the sparse graphs are only able to display a low number of edges (inter-nodal links), thus, making the possible solutions subsequently too low. Additionally, the achieved results demonstrate well that the BPSO turns out to record a noticeably higher performance as compared to the GA recorded scores. For instance, when the HQ value is equal to 50%, an improvement by 2% turns out to be registered.

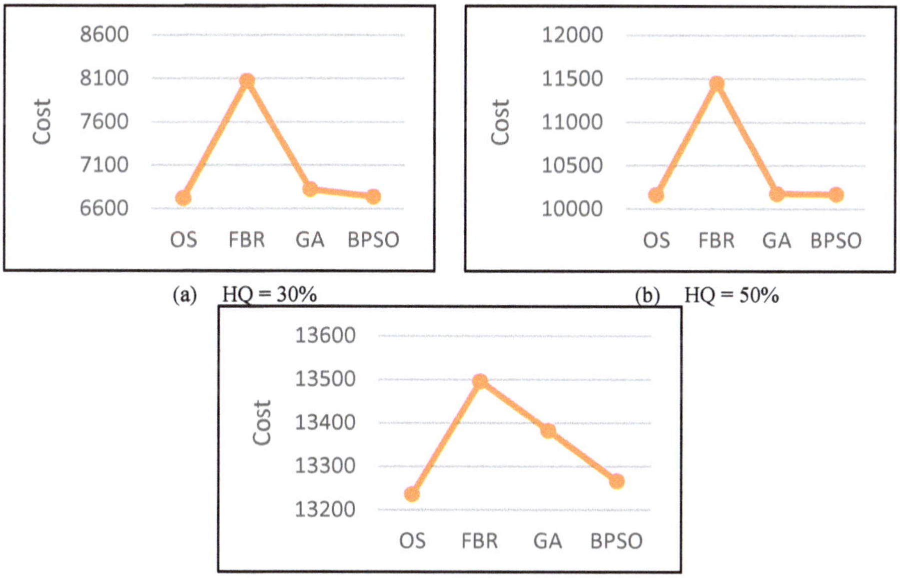

(a) HQ = 30% (b) HQ = 50%

(c) HQ = 70%

Fig. 5. Cost comparison respectively involving the BPSO, GA, FBR and OS in terms of graph G10

Figure 5 depicts the cost ranges as reached on a full graph basis, highlighting that the BPSO scored costs are discovered to be remarkably close to the optimal solutions. Similarly, the BPSO turns out to record rather excellent results, exceeding even the FBR scored ones, for the entirety of cases, exhibiting a different rate of even 50%. Noteworthy, also, is that the GA appears to score cost ranges that are too close to those registered via BPSO.

As to the Fig. 6, it highlights the costs recorded on a dense graph, at a density range level of 50%. The reached results indicate well that the BPSO turns out to perform better than the FBR with respect to most of the HQ variation cases. Noteworthy, also, is that the GA tends to score cost levels that are somewhat too close, sometimes even a bit higher, than those recorded by the BPSO.

An analysis of the Figs. 4, 5 and 6 reveals well that, for any graph type, the BPSO turns out to record more effective results than the FBR, particularly with regard to the dense graphs, wherein the BPSO seems to demonstrate a highly excellent behaviour.

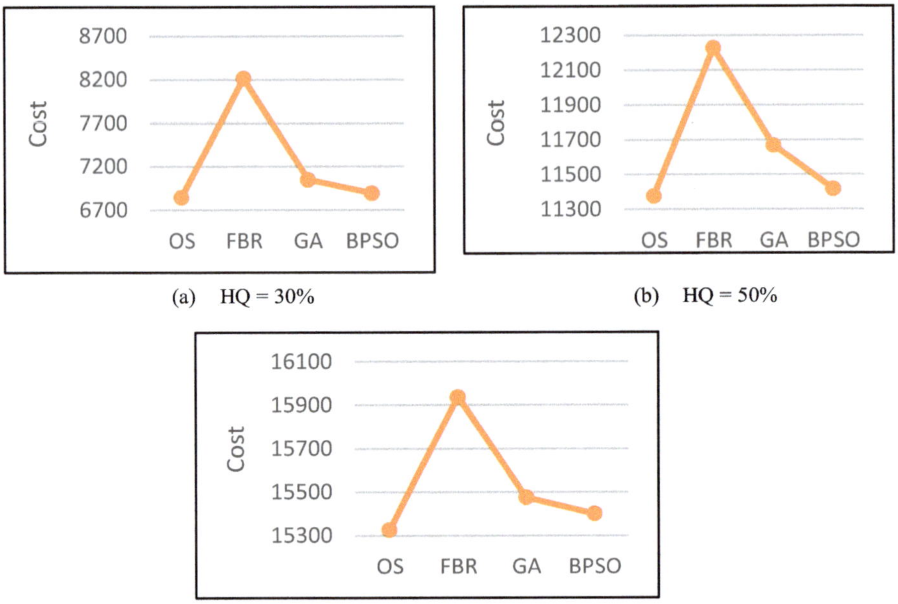

(a) HQ = 30%

(b) HQ = 50%

(c) HQ = 70%

Fig. 6. Cost comparison involving respectively the BPSO, GA, FBR and OS as based on graph G5

Besides, the BPSO displays a better result than that scored via GA, although there are cases wherein the results scored turn out to be remarkably close.

7 Conclusion and Potential Work Perspective

This research is focused on treating the NP-hard problem, relevant to the placement of Service-Based Applications (SBA), as deployed via hybrid cloud contexts.

In this context, a novel BPSO based approach has been put forward, whereby the SBAs placement process on a hybrid cloud platform could be highly optimized. For the sake of an effective evaluation of the advanced approach, 2000 experiments have been administered, relying on the IBM provided data set benchmark. The reached experimental results appeared to reveal well that the proposed approach turns out to display a remarkably effective behaviour, recording formidable scores, as compared to both of the state-of-the-art FBR and GA algorithms.

Actually, the remarkable work achieved turns out to be very promising, and several potential study venues are under consideration. As a potential work perspective, for instance, one could well envisage extending the approach even further so that the placement related problem could be resolved within an online framework design.

Acknowledgments. The research leading to the recorded results achievements has received funding from the Ministry of Higher Education and Scientific Research of Tunisia, under grant agreement number: LR11ES48.

References

1. Sahni, S.: Computationally related problems. SIAM J. Comput. **3**, 262–279 (1974)
2. Yusoh, Z., Tang, M.: A penalty-based genetic algorithm for the composite SaaS placement problem in the cloud. In: 2010 IEEE Congress on Evolutionary Computation (CEC), pp. 1–8, July 2010
3. Wada, H., Suzuki, J., Yamano, Y., Oba, K.: E3: A multiobjective optimization framework for SLA-aware service composition. IEEE Trans. Serv. Comput. **5**(3), 358–372 (2012)
4. Li, W., Zhong, Y., Wang, X., Cao, Y.: Resource virtualization and service selection in cloud logistics. J. Netw. Comput. Appl. **36**(6), 1696–1704 (2013)
5. Klein, A., Ishikawa, F., Honiden, S.: SanGA: A self-adaptive network-aware approach to service composition. IEEE Trans. Serv. Comput. **7**(3), 452–464 (2014)
6. Yusoh, Z., Tang, M.: A penalty-based grouping genetic algorithm for multiple composite saas components clustering in cloud. In: 2012 IEEE International Conference on Systems, Man, and Cybernetics (SMC), pp. 1396–1401, October 2012
7. Z. Yusoh and M. Tang, "Composite SaaS placement and resource optimization in cloud computing using evolutionary algorithms," in 2012 IEEE 5th International Conference on Cloud Computing (CLOUD), pp. 590–597, June 2012.
8. Deb,] K., Agrawal, S., Pratap, A., Meyarivan, T.: A fast and elitist multiobjective genetic algorithm: NSGA-II. IEEE Trans. Evol. Comput. **6**(2), 182–197 (2002)
9. Klein, A., Ishikawa, F., Honiden, S.: Towards network-aware service composition in the cloud. In: Proceedings of the 21st International Conference on World Wide Web, WWW 2012, New York, pp. 959–968, ACM (2012)
10. Kaviani, N., Wohlstadter, E., Lea, R.: Manticore: a framework for partitioning software services for hybrid cloud. In: Proceedings of the 2012 IEEE 4th International Conference on Cloud Computing Technology and Science (CloudCom), CLOUDCOM 2012. Washington, DC, pp. 333–340, IEEE Computer Society (2012)
11. Ben Charrada, F., Tata, S.: An efficient algorithm for the bursting of service-based applications in hybrid clouds. IEEE Trans. Serv. Comput. **9**(3), 357–367 (2016)
12. Abbes, W., Kechaou, Z., Alimi, A.M.: A new placement optimization approach in hybrid cloud based on genetic algorithm. In: IEEE International Conference on e-Business Engineering (ICEBE), Macau (2016)
13. Lee, Y. C., Lian, B.: Cloud bursting scheduler for cost efficiency. In: 2017 IEEE 10th International Conference on Cloud Computing (CLOUD), pp. 774–777 (2017)
14. Barthwal, V., Rauthan, M.M.S.: AntPu: a meta-heuristic approach for energy-efficient and SLA aware management of virtual machines in cloud computing. Memet. Comput. **13**(1), 91–110 (2021). https://doi.org/10.1007/s12293-020-00320-7
15. Lucas-Simarro, J., Moreno-Vozmediano, R., Montero, R., Llorente, I.: Dynamic placement of virtual machines for cost optimization in multicloud environments. In: Proceedings of the 2011 International Conference on High Performance Computing & Simulation (HPCS), pp. 1–7 (2011)
16. Phani Praveen, S., Tulasi, U., Ajay Krishna Teja, K.: A cost efficient resource provisioning approach using virtual machine placement. Int. J. Comput. Sci. Inf. Technol. **5**(2), 2365–2368 (2014)
17. Business Process Incubator, June 2021. https://www.businessprocessincubator.com/category/type/templates/
18. Dijkman, R., Hofstetter, J., Koehler, J. (eds.): BPMN 2011. LNBIP, vol. 95. Springer, Heidelberg (2011). https://doi.org/10.1007/978-3-642-25160-3
19. Kennedy, J., Eberhart, R.C.: A discrete binary version of the particle swarm algorithm. In: IEEE International Conference on Systems, Man, and Cybernetics (1997)

20. Fahland, D., Favre, C., Koehler, J., Lohmann, N., Volzer, H., Wolf, K.: Analysis on demand: Instantaneous soundness checking of industrial business process models. Data Knowl. Eng. **70**(5), 448–466 (2011)
21. ILOG SA, ILOG CPLEX 12, User's Manual (2017). https://www-01.ibm.com/software/commerce/optimization/cplex-optimizer/

Profile Deviation Analysis of Global Firms' Working Capital Management in the Automotive Industry During the Financial Crisis and Recovery Periods

Keontaek Oh and DaeSoo Kim(✉)

Korea University Business School, Seoul, Korea
kimd@korea.ac.kr

Abstract. In uncertain industrial environment with economic downturn, firms are faced with various risks such as financing and liquidity, supply chain disruption, and demand shock, etc., hence the need for effective working capital management (WCM). Yet in extant studies, little is explored on the "adequate" levels of the cash conversion cycle (CCC) and its elements (days of inventory outstanding, days of accounts receivable outstanding, and days of accounts payable outstanding), the typical measures of WCM. Therefore, this study aims to identify an empirically ideal profile of the CCC and its elements of the top performance group in terms of the average net sales growth rate during the global financial crisis and recovery periods, using the data of the Forbes Global 2000 ranking firms in the automotive industry. Then, based on the concept of profile deviation, we analyze its profile difference with other groups in each period, using non-parametric test. The findings provide theoretical and practical implications on effective WCM.

Keywords: Working capital management · Cash conversion cycle · Financial crisis and recovery · Profile deviation · Non-parametric test · Forbes Global 2000

1 Introduction

In today's highly competitive and uncertain industrial environment with economic downturn, caused by global financial crisis, trade protectionism, disasters and pandemic, etc., firms are confronted with various risks, including financing and liquidity, supply chain disruption, and demand shock. So, effective working capital management (WCM) has become more important today than before [21, 22, 28]. It is because effective WCM can reduce financial risk and increase profitability through liquid asset management [4], maintain stable cash flow and financial position, create potential value [7, 21, 46], and influence the relationship with customers and suppliers [2].

Extant WCM studies have frequently used the cash conversion cycle (CCC) as a key evaluation metric, typically defined as the sum of the days of inventory outstanding (DIO) and the days of accounts receivable outstanding (DRO) minus the days of accounts

This study was partially supported by the Korea University Business School Research Grant.

R. Lee (Ed.): ICIS 2021, SCI 1003, pp. 27–37, 2022.
https://doi.org/10.1007/978-3-030-90528-6_3

payable outstanding (DPO) [5, 13, 17]. Most of these studies have investigated the causal relationship between the CCC and/or its elements (i.e., DIO, DRO, DPO) and various financial performance measures of firms with different size in various industries and countries for different periods [e.g., 15, 20, 25, 28, 31, 45]. Other studies have examined the CCC and other WCM implementation policies [19, 23]. Yet, little is discovered about the fundamental inquiry in both practice and academia as to *what are the "adequate" levels or ranges of the CCC and its elements (i.e., DIO, DRO, and DPO) under different economic conditions.*

Therefore, this study aims to identify the characteristics of an empirically ideal profile of the top performance group's CCC and its elements during the global financial crisis and recovery periods and compare its profile with other groups', based on the concept of profile deviation [40] used in corporate strategy research. The primary purpose is to build a theoretical foundation of effective WCM and provide practical insights on the "adequate" levels or ranges of the CCC, DIO, DRO, and DPO. To do so, using the Forbes Global 2000 ranking in 2016, we identified 42 firms in the automotive industry, as the shock could strongly influence their business partners and the economy as well [16]. For the analysis, we used the financial data compiled from Data Stream during the global financial crisis (2008–2011) and recovery (2012–2015) periods and employed non-parametric test [34] to compare an ideal profile of the top performance group in terms of the average net sales growth rate with other groups' in each period.

This study provides profound theoretical and practical contributions. From a theoretical perspective, it lays out the foundation for analyzing an empirically ideal profile of the CCC and its elements under different economic conditions. From a practical standpoint, the empirically-driven "adequate" levels of the CCC and its elements provide the insights into how to effectively manage not only working capital but also customer and supplier relationships for sales growth and liquidity.

This paper is organized as follows. Section 2 reviews extant studies and presents the theoretical background of this study. Section 3 describes research design, including data collection and research method. Section 4 presents and discusses analysis results. And Sect. 5 concludes with the summary, limitations and future research directions.

2 Literature Review and Theoretical Background

Extant studies of working capital management (WCM) have mostly used cash conversion cycle (CCC) as its primary evaluation metric, which is typically defined as the sum of the days of inventory outstanding (DIO) and the days of accounts receivable outstanding (DRO) minus the days of accounts payable outstanding (DPO) [5, 13, 17]. Most of the studies have investigated the causal relationship between the CCC and/or its elements and various financial performance measures (e.g., ROS, ROA, gross or net operating income, Tobin's Q). Those studies were conducted for firms with different size (small and medium-sized, large enterprises) in various industries (e.g., agriculture, service, manufacturing, construction, metal, wholesale, textile, petrochemical, finance industries, etc.) and countries (USA, Saudi Arabia, Japan, Taiwan, Sweden, Belgium, Greece, Vietnam, etc.) [3, 4, 6, 8, 13, 18, 20, 21, 25, 26, 30, 35, 43, 45]. In most of them, the CCC and its elements (DIO, DRO, and DPO) have shown negative and no relationship with diverse profitability measures, respectively, while in a few studies its elements have indicated negative relationship; see [28, 29] for a comprehensive review.

A few studies, like the present study, have examined the periods associated with the global financial crisis. Ramiah et al. [31] studied the CCC elements of Australian firms in energy, financial, materials, information and other industries, and found that during the global financial crisis period they implemented conservative WCM policies by reinforcing the credit and reducing the expenditure, inventory and CCC. Haron & Nomran [15] examined Malaysian firms before, during, and after the global financial crisis, and discovered that the CCC was negatively correlated with ROA in all three periods and with sales growth before and after the crisis, while the effect was mixed with other financial measures. Oh et al. [28] investigated the Forbes Global 2000 firms in the automotive and electrical-electronics industries during and after the global financial crisis, and revealed that the DIO had a negative effect on ROS in both industries in both periods, while the DRO had a negative effect in the automotive industry during the crisis period and the DPO did not show any effect.

Other studies have examined the CCC and other WCM implementation policies. Koury et al. [19] analyzed various CCC policies of small and large firms in the US, Canada and Australia, using multi-year surveys, and showed various factors affecting the implementation of CCC policies. Michalski [23] conducted case studies of the firms' credit and asset management policies, and found that the firms' credit management affected the DRO, and in turn influenced the management cost of the CCC.

The literature review reveals that little research has been done on exploring the "adequate" levels of global top firms' CCC and its element under different economic conditions. Therefore, this study aims to identify an empirically ideal profile of the top performance group and compare its profile with other groups' during the global financial crisis and recovery periods, based on the concept of profile deviation [40]. The purpose is to provide a theoretical foundation of effective WCM and practical insights. In theory building, the concept of fit has played an important role in the areas of organization theory and corporate strategy [1, 5, 10, 24, 36–38, 41]. Fit as profile deviation is the degree of adherence to an externally specified (theoretical or empirically ideal) profile [40], similarly as in pattern analysis [38]. Developing the ideal profile could lead to a benchmark for strategic planning, e.g., strategic resource deployments [42]. As such, identifying an empirically ideal profile of the global top performers' CCC and its elements could help practitioners better understand how much they are deviated from the ideal and hence establish more effective WCM policies.

3 Research Design

This study investigates an empirically ideal profile of the global firms' cash conversion cycle (CCC) and its elements (DIO, DRO and DPO) in the automotive industry during the global financial crisis (2008–2011) and recovery (2012–2015) periods, based on the concept of profile deviation [40]. To do so, using the Forbes Global 2000 ranking (based on the equal weight of sales, profit, assets, and market value) in 2016, we identified the global top 42 firms (automotive, truck, and part manufacturers) in the automotive industry [28] listed in Table 1. The choice of the automotive industry was due to the strong impact of the financial shock on business partners and economy [16].

For the analysis, we compiled the panel data for inventories, accounts receivable, accounts payable, net sales, cost of goods sold and purchase cost from the World Scope

Table 1. Global top 42 firms in the automotive industry

Toyota Motor, Volkswagen Group, Daimler, Ford Motor, BMW Group, Honda Motor, Nissan Motor, Hyundai Motor, SAIC Motor, Renault, Continental, Denso, Bridgestone, Tata Motors, KIA Motors, Hyundai MOBIS, Fuji Heavy Industries, Peugeot, Toyota Industries, Michelin Group, Magna International, Johnson Controls, Suzuki Motor, Mazda Motor, Aisin Seiki, Sumitomo Electric, Isuzu Motors, BYD, Chongqing Changan Auto, Great Wall Motor, Mahindra & Mahindra, Lear, Goodyear, Mitsubishi Motors, Autoliv, BorgWarner, LKQ, GKN, Visteon, JTEKT, Sumitomo Rubber, Toyota Boshoku

database of Data Stream for the global financial crisis and recovery periods [28]. The panel data for research variables were obtained, based on the definition in Table 2.

Table 2. Definition of research variables

DIO (days of inventory outstanding) = (inventory / cost of goods sold) x 365
DRO (days of accounts receivable outstanding) = (accounts receivable / net sales) x 365
DPO (days of accounts payable outstanding) = (accounts payable / purchase cost) x 365
CCC (cash conversion cycle) = DIO + DRO – DPO
NSGR (net sales growth rate): based on geometric mean in each period

Next, we divided the firms into the top performance group and other groups based on the average net sales growth rate (NSGR) in each period, with an equal ten percent interval except for the first and the last group. As a result, three groups were formed in each period, as shown in the next section.

Then, based on the concept of profile deviation [40], we identified an empirically ideal profile of the top performance group's CCC, DIO, DRO and DPO, and compared its profile with other groups', using non-parametric test in SPSS 21 [27, 34], instead of profile deviation scores as in [5, 40, 42]. It is because our focus is not on examining the impact of the profile deviation on dependent variables, but on discovering profile differences of other groups from the top performance (ideal) group. Further, the choice of non-parametric test (i.e., Mann-Whitney *U*, Wilcoxon W, and Z test) was due to non-normality of most of the group samples from Shapiro-Wilk test results [32, 44].

4 Analysis and Discussion

This section presents and discusses the analysis results of an empirically ideal profile of the top performance group's (G1) CCC, DIO, DRO and DPO, and other groups' (G2~G3) profile differences in the automotive industry during the global financial crisis (2008–2011) and recovery (2012–2015) periods.

4.1 Profile Deviation During the Global Financial Crisis Period (2008–2011)

Table 3 shows the profiles (descriptive statistics) of three groups' CCC, DIO, DRO, and DPO, along with net sales growth rate during the global financial crisis period. The top performance group G1's profile is the empirically ideal profile in terms of the average net sales growth rate (NSGR = 25.2%), not the average net sales, with its mean CCC = 60.0, DIO = 36.8, DRO = 87.2, and DPO = 64.0 days. It should be noted that the average net sales were the highest in G2. For G1's profile compared with G2 and G3, its mean CCC was in the low side but not the lowest and its mean DIO was the lowest, while both its mean DRO and DPO were notably the highest. Compared with the total sample, its mean CCC and DIO were shorter, while both its mean DRO and DPO were longer.

Table 3. Group profiles (descriptive statistics) during the financial crisis period

Group	CCC	DIO	DRO	DPO	NSGR	Net sales ($10 Million)
G1 (11)	60.0, 59.6, 1.8, 167.6	36.8, 13.9, 16.1, 60.4	87.2, 46.1, 28.5, 185.0	64.0, 30.3, 23.6, 127.3	25.2, 10.3, 10.3, 42.3	505.5, 1083.0, 0.2, 3428.6
G2 (12)	81.1, 42.6, 24.9, 151.1	49.1, 16.8, 16.5, 78.8	76.0, 34.2, 43.3, 164.8	44.0, 10.8, 23.3, 57.2	4.0, 2.6, 0.5, 9.9	774.1, 2492.5, 0.5, 9039.6
G3 (19)	58.1, 38.8, 10.1, 140.4	38.1, 13.7, 9.7, 71.6	66.0, 36.6, 23.3, 162.0	46.0, 19.0, 14.1, 105.2	−4.0, 2.0, −8.0, −0.2	334.0, 496.9, 0.7, 2119.0
Total (42)	65.1, 49.2, 1.8, 167.6	40.9, 16.9, 9.7, 78.8	74.4, 41.3, 23.3, 185.0	50.1, 23.5, 14.1, 127.3	6.1, 1.8, −8.0, 42.3	537.9, 1357.5, 0.2, 9039.6

Notes) CCC = cash conversion cycle; DIO = days of inventory outstanding; DRO = days of accounts receivable outstanding; DPO = days of accounts payable outstanding; NSGR = average net sales growth rate (G1: 10%~, G2: 0~<10%, G3: ~<0%); Numbers in each cell: Mean, Standard deviation, Min, Max; Numbers under each group: Sample size.

Table 4 summarizes the non-parametric test results during the financial crisis period. It showed that there are statistically significant differences in DIO between G1 and G2 ($p < 0.1$) and between G2 and G3 ($p < 0.1$), and in DPO between G1 and G2 ($p < 0.1$) and between G1 and G3 ($p < 0.05$). It also revealed that there are no significant differences among the groups in DRO and CCC. The profile differences among the groups may be attributed to the differences in production methods, inventory management strategies, credit and liquidity management strategies, and customer or supplier relationship management strategies [9, 39].

From Tables 3 and 4 together, we can excavate some interesting implications on planning and implementing WCM strategies, when firms are confronted with global financial crisis. First, the top performance group G1 in terms of NSGR showed the significant but weak difference ($p < 0.1$) in DIO and DPO with G2 whose average net sales were higher than G1. Interestingly, G1 had the lowest DIO and the highest DPO, with the highest DRO, while G2 had the highest DIO and the lowest DPO, opposite to

G1, with the high DRO. From this, we can infer that in the financial crisis period G1 had focused on both sales growth and liquidity, given the lowest DIO and the highest DRO and DPO. In contrast, G2 had focused more on net sales increase and less on liquidity, given the highest DIO and the high DRO but the lowest DPO. This reasoning comes from the general notion that the increase in DIO and DRO would lead to the increase in sales but the decrease in liquidity, while the increase in DPO would only lead to the decrease in liquidity [29]. Second, the lowest performance group G3 with negative net sales growth showed the lowest CCC with the low DIO, the lowest DRO and the low DPO, focusing mostly on liquidity and not on sales growth [12]. Therefore, we can infer that the lowest CCC with the low DIO and the lowest DRO for better liquidity could lead to negative sales growth [30].

Table 4. Non-parametric test results during the financial crisis period

CCC	DIO	DRO	DPO
-	G1, G2 (0.069*), (36.00, 102.00, −1.85)	-	G1, G2 (0.091*), (38.00, 116.00, −1.72)
-	G2, G3 (0.053*), (66.00, 256.00, −1.95)	-	G1, G3 (0.047**), (58.00, 248.00, −2.00)

Notes) * $p < 0.1$, ** $p < 0.05$, *** $p < 0.01$; CCC = cash conversion cycle; DIO = days of inventory outstanding; DRO = days of accounts receivable outstanding; DPO = days of accounts payable outstanding; G = Group divided by average net sales growth rate (G1: 10%~, G2: 0~<10%, G3: ~<0%); Numbers in each cell: (*p*-value), (Mann-Whitney *U*, Wilcoxon W, Z).

4.2 Profile Deviation During the Global Financial Recovery Period (2012–2015)

Table 5 shows the profiles (descriptive statistics) of three groups' CCC, DIO, DRO, and DPO, along with net sales growth rate during the global financial recovery period. The top performance group G1's profile is the empirically ideal profile based on the average net sales growth rate (NSGR = 17.4%), not the average net sales, with its mean CCC = 45.1, DIO = 37.5, DRO = 67.8, and DPO = 60.1 days. For G1's profile compared with G2 and G3, its mean CCC is the lowest, with its DIO and DRO being almost the lowest, but its DPO being about in the middle. Compared with the total sample, its mean CCC, DIO, and DRO were shorter, while its mean DPO was longer.

The non-parametric test results during the financial recovery period revealed no statistically significant differences among the groups in CCC, DIO, DRO, and DPO. This result may be ascribable to too small sample size of G3. Nonetheless, from Table 5, we can observe interesting patterns of the profiles among the groups. First, the top performance group G1 in terms of NSGR had the lowest DIO and DRO with the middle level of DPO, yielding the lowest CCC, while G2 with the notably largest net sales and the modest NSGR had the lowest DIO and the highest DRO with the lowest DPO, resulting in the highest CCC. From this, we can infer that in the financial recovery period G1 had focused mostly on internal operations efficiency and liquidity, given the lowest DIO, DRO and CCC. Differently, G2 had concentrated more on not only internal operations efficiency but also better customer and supplier relationships and less on liquidity, given the lowest DIO and DPO and the highest DRO and CCC. Second, the

Table 5. Group profiles (descriptive statistics) during the financial recovery period

Group	CCC	DIO	DRO	DPO	NSGR	Net sales ($10 Million)
G1 (10)	45.1, 39.7, 7.5, 112.2	37.5, 12.1, 20.5, 62.3	67.8, 40.3, 20.8, 132	60.1, 18.4, 34.4, 86.7	17.4, 5.7, 10, 26.2	182.9, 308.9, 1074.9, 0.5
G2 (28)	74.3, 50.8, 9.8, 213.2	37.5, 14.2, 12.9, 75.2	82.7, 47.7, 26, 242.5	45.8, 12.2, 19.2, 71.4	6.0, 2.4, 0.5, 9.9	818.3, 1898.7, 0.6, 8824.8
G3 (4)	59.4, 26.8, 24.3, 85.9	63.9, 9.6 54.4, 78.5	67.8, 33.7, 28.3,116.5	72.2, 51.6 21.8, 158.5	−1.6, 4.3, −20.3, 3.8	2.2, 1.9, 0.6, 5.4
Total (42)	65.9, 50.3, 7.5, 213.2	39.9, 15.7, 12.9, 78.5	77.6, 47.2, 20.8, 242.5	51.7, 23.0, 19.2, 158.5	6.8, 8.6 −20.3, 26.2	589.3, 1591.7 0.5, 8824.8

Notes) CCC = cash conversion cycle; DIO = days of inventory outstanding; DRO = days of accounts receivable outstanding; DPO = days of accounts payable outstanding; NSGR = average net sales growth rate (G1: 10%~, G2: 0~<10%, G3: ~<0%); Numbers in each cell: Mean, Standard deviation, Min, Max; Numbers under each group: Sample size.

lowest performance group G3 with negative net sales growth had the highest DIO, the lowest DRO, and the highest DPO, showing relatively inconsistent strategies for sales growth or liquidity as well as less concerns on internal operations efficiency and customer and supplier relationships. Therefore, we can infer that inconsistent WCM strategies of reducing working capital at the expense of its value chain partners could lead to negative sales growth [22].

4.3 Comparative Period Analysis

The non-parametric test results for each group between the global financial crisis and recovery periods revealed no statistically significant differences in CCC, DIO, DRO, and DPO. Nonetheless, comparing Table 3 with Table 5 for the entire group and each group by periods, we also find interesting implications.

First, for the entire group of 42 global top firms in the automotive industry based on the Forbes Global 2000 ranking, there were only small changes from the crisis period to the recovery period in the mean CCC (65.1 to 65.9), DIO (40.9 to 39.9), DRO (74.4 to 77.6), and DPO (50.1 to 51.7). Further, the number of firms with positive NSGR increased from 23 to 38. Second, for the top performance group G1 in terms of NSGR, its mean CCC was reduced somewhat from 60.0 to 45.1, with the small increase in DIO (36.8 to 37.5), the modest decrease in DRO (87.2 to 67.8), and the small decrease in DPO (64.0 to 60.1). And both average NSGR and net sales were decreased. Therefore, G1's strategies in the crisis period focused more on improving sales growth and customer relationship, whereas those in the recovery period concentrated more on liquidity and supplier relationship rather than improving net sales growth. Third, for G2 with the largest net sales, its mean CCC was reduced somewhat from 81.1 to 74.3, with the modest decrease in DIO (49.1 to 37.5), the modest increase in DRO (76.0 to 82.7), and the small increase in DPO (44.0 to 45.8). And both average NSGR and net sales were increased. Therefore, G2's strategies in the recovery period, compared with the crisis period, focused more on liquidity, internal operations efficiency, and net sales increase with better customer relationship. And fourth, for the lowest performance group G3 with negative NSGR, its mean CCC was increased a little from 58.1 to 59.4, with the rather sharp increase in DIO (38.1 to 63.9), the small increase in DRO (66.0 to 67.8), and the rather sharp increase in DPO (46.0 to 72.2). The number of firms in this group was decreased sharply in the recovery period, as expected. While G3's strategies in the crisis period were geared toward too much on liquidity and not on sales growth, those in the recovery period were rather problematic with too high DIO and DPO, thereby deteriorating internal operations efficiency and supplier relationship.

5 Conclusion

This study investigated the Forbes 2000 ranking firms in the automotive industry to identify an empirically ideal profile (CCC, DIO, DRO, and DPO) of the top performance group in terms of the average net sales growth rate during the global financial crisis and recovery periods and to analyze its profile difference with other groups', using non-parametric test, based on the concept of profile deviation. The analysis results provided both theoretical and practical implications on the "adequate" levels of CCC, DIO, DRO, and DPO.

First, during the crisis period, the top performance group G1 in terms of net sales growth rate (NSGR) focused on both sales growth and liquidity, with the lowest DIO and the highest DRO and DPO. In contrast, G2 with the largest average net sales focused more on sales increase and less on liquidity, with the highest DIO and the high DRO but the lowest DPO. Moreover, the lowest CCC with the low DIO and the lowest DRO for better liquidity could lead to negative sales growth.

Second, during the recovery period, the top performance group G1 in terms of NSGR focused mostly on internal operations efficiency and liquidity, given the lowest DIO, DRO and CCC. Differently, G2 with the largest average net sales concentrated more on not only internal operations efficiency but also better customer and supplier relationships and less on liquidity, given the lowest DIO and DPO and the highest DRO and CCC. Moreover, inconsistent WCM strategies of reducing working capital at the expense of its value chain partners could lead to negative sales growth.

And third, from the crisis period to the recovery period, the global top firms showed only small changes in the mean CCC, DIO, DRO, and DPO, while each group's changes were bigger. G1 showed the modest decrease in CCC, with the small increase in DIO, the modest decrease in DRO, and the small decrease in DPO, with its strategic focus shifted from sales growth and customer relationship to liquidity and supplier relationship. G2 also showed the modest decrease in CCC, but differently with the modest decrease in DIO, the modest increase in DRO, and the small increase in DPO, with its strategic focus shifted from mainly sales growth and customer relationship to liquidity, internal operations efficiency, and sales increase with better customer relationship. Moreover, too much focus on liquidity and not on sales growth in the crisis period and too sharp increase in DIO and DPO in the recovery period could lead to poor WCM.

Despite its contributions, this study is not free from its limitations in terms of generalizability, which can serve as future research directions. First, future study can replicate the present study for different crisis and recovery periods. Second, future study can also replicate the present study for other industries. Such studies would become important stepping-stones toward building a theory on WCM, thereby proving more concrete and fruitful insights to the practice.

References

1. Aldrich, H.E.: 1979 Organizations and Environments. Prentice-Hall (1979)
2. Al-Shubiri, F.N., Aburumman, N.M.: The relationship between cash conversion cycle and financial characteristics of industrial sectors: an empirical study. Investment Manag. Financ. Innovations **10**(4), 95–102 (2013)
3. Appuhami, B.: The impact of firms' capital expenditure on working capital management: an empirical study across industries in Thailand. Int. Manag. Rev. **4**(1), 8 (2008)
4. Deloof, M.: Does working capital management affect profitability of Belgian firms? J. Bus. Financ. Acc. **30**(3–4), 573–588 (2003)
5. Drazin, R., Van de Ven, A.H.: Alternative forms of fit in contingency theory. Adm. Sci. Q. **30**(4), 514–539 (1985)
6. Eljelly, A.M.: Liquidity-profitability tradeoff: an empirical investigation in an emerging market. Int. J. Commer. Manag. **14**(2), 48–62 (2004)
7. Fazzari, S.M., Petersen, B.C.: Working capital and fixed investment: new evidence on financing constraints. Rand J. Econ. **24**(3), 328–342 (1993)
8. Ferrando, A., Mulier, K.: Do firms use the trade credit channel to manage growth? J. Bank. Financ. **37**(8), 3035–3046 (2013)
9. Fisher, M.L., Hammond, J.H., Obermeyer, W.R., Raman, A.: Making supply meet demand in an uncertain world. Harv. Bus. Rev. **72**, 83 (1994)
10. Fry, L.W., Smith, D.A.: Congruence, contingency, and theory building. Acad. Manag. Rev. **12**(1), 117–132 (1987)
11. García-Teruel, P.J., Martínez-Solano, P.: Effects of working capital management on sme profitability. Int. J. Manag. Financ. **3**(2), 164–177 (2007)
12. Gentry, J.A., Vaidyanathan, R., Lee, H.W.: A Weighted cash conversion cycle. Financ. Manag. **19**(1), 90–99 (1990)
13. Gitman, L.J.: Estimating corporate liquidity requirements: a simplified approach. Financ. Rev. **9**(1), 79–88 (1974)
14. Habib, A., Bhuiyan, B.U., Islam, A.: Financial distress, earnings management and market pricing of accruals during the global financial crisis. Manag. Financ. **39**(2), 155–180 (2013)
15. Haron, R., Nomran, N.M.: Determinants of working capital management before, during, and after the global financial crisis of 2008: evidence from Malaysia. J. Develop. Areas. **50**(5), 461–468 (2016)
16. Haugh, D., Mourougane, A., Chatal, O.: The Automobile Industry in and beyond the Crisis. In: OECD Economics Department Working Papers, **745**, 1–37 (2010)
17. Keown, A.J., Martin, J.D., Petty, J.W., Scott Jr, D.F.: Foundations of Finance, 9th ed. Pearson, London (2016)
18. Khan, W., Ellahi, A.R., Mehmood, B.: Impact of financial supply chain management (fscm) on financial performance of textile sector in Pakistan: empirical evidence from generalized method of moments. Sci. Int. **28**(1), 1563–1568 (2016)
19. Koury, N.T., Smith, K.V., MacKay, P.I.: comparing working capital practices in Canada, the United States, and Australia: a note. purdue cyber working papers, 1–13 (1998)
20. Lazaridis, I., Tryfonidis, D.: relationship between working capital management and profitability of listed companies in the athens stock exchange. J. Financ. Manag. Anal. **19**(1), 26–35 (2006)

21. Le, B.: Working capital management and firm's valuation, profitability and risk. Int. J. Manag. Financ. **15**(2), 191–204 (2019)
22. Lind, L., Pirttilä, M., Viskari, S., Schupp, F., Kärri, T.: Working capital management in the automotive industry: financial value chain analysis. J. Purch. Supply Manag. **18**(2), 92–100 (2012)
23. Michalski, G.: Portfolio management approach in trade credit decision making. Romanian J. Econ. Forecast. **3**, 42–53 (2007)
24. Miles, R.E., Snow, C.C., Meyer, A.D., Coleman, H.J., Jr.: Organizational strategy, structure, and process. Acad. Manag. Rev. **3**(3), 546–562 (1978)
25. Muscettola, M.: Cash conversion cycle and firm's profitability: an empirical analysis on a sample of 4,226 manufacturing smes of italy. Int. J. Bus. Manag. **9**(5), 25–35 (2014)
26. Nguyen, A.H., Pham, H.T., Nguyen, H.T.: Impact of working capital management on firm's profitability: empirical evidence from Vietnam. J. Asian Financ., Econ., Bus. **7**(3), 115–125 (2020)
27. Nie, N.H., Bent, D.H., Hull, C.H.: SPSS: Statistical Package for the Social Sciences. McGraw-Hill, New York (1975)
28. Oh, K.T., Lee, D.H., Kim, D., Jeong, E.B.: Panel data analysis of the relationship between cash conversion cycle elements and profitability in the global automotive and electrical-electronics industries. Global Bus. Admin. Rev. **17**(2), 212–236 (2020)
29. Prasad, P., Narayanasamy, S., Paul, S., Chattopadhyay, S., Saravanan, P.: Review of literature on working capital management and future research agenda. J. Econ. Surv. **33**(3), 827–861 (2019)
30. Raheman, A., Nasr, M.: Working capital management and profitability-case of Pakistani firms. Int. Rev. Bus. Res. Papers. **3**(1), 279–300 (2007)
31. Ramiah, V., Zhao, Y., Moosa, I.: Working Capital Management during the global financial crisis: the Australian experience. Qual. Res. Financ. Markets. **6**(3), 332–351 (2014)
32. Razali, N.M., Wah, Y.B.: Power comparisons of Shapiro-Wilk, Kolmogorov-Smirnov, Lilliefors and Anderson-Darling tests. J. Stat. Model. Anal. **2**(1), 21–33 (2011)
33. Richards, V.D., Laughlin, E.J.: A cash conversion cycle approach to liquidity analysis. Financ. Manag. **9**(1), 32–38 (1980)
34. Ruxton, G.D.: The Unequal variance t-test is an underused alternative to student's t-test and the Mann-Whitney u test. Behav. Ecol. **17**(4), 688–690 (2006)
35. Shin, H.H., Soenen, H.L.: Efficiency of working capital and corporate profitability. Financ. Pract. Educ. **8**(2), 37–45 (1998)
36. Snow, C.C., Miles, R.E.: The role of strategy in the development of a general theory of organizations. Adv. Strateg. Manag. **2**(1), 213–259 (1983)
37. Thompson, J.D.: Organizations in Action. McGraw-Hill, London (1967)
38. Van de Ven, A.H., Drazin, R.: The Concept of Fit in Contingency Theory. Strategic Management Research Center. 1–74, (1984)
39. Van Wezel, W., Van Donk, D.P., Gaalman, G.: The planning flexibility bottleneck in food processing industries. J. Oper. Manag. **24**(3), 287–300 (2006)
40. Venkatraman, N.: The concept of fit in strategy research: toward verbal and statistical correspondence. Acad. Manag. Rev. **14**(3), 423–444 (1989)
41. Venkatraman, N., Camillus, J.C.: Exploring the concept of "fit" in strategic management. Acad. Manag. Rev. **9**(3), 513–525 (1984)
42. Venkatraman, N., Prescott, J.E.: Environment-strategy coalignment: an empirical test of its performance implications. Strateg. Manag. J. **11**(1), 1–23 (1990)
43. Wang, Y.J.: Liquidity management, operating performance, and corporate value: evidence from Japan and Taiwan. J. Multinatl. Financ. Manag. **12**(2), 159–169 (2002)
44. Wonnacott, T.H., Wonnacott, R.J.: Introductory Statistics for Business and Economics. Wiley. Hoboken (1990)

45. Yazdanfar, D., Öhman, P.: The impact of cash conversion cycle on firm profitability. Int. J. Manag. Financ. **10**(4), 442–452 (2014)
46. Zariyawati, M.A., Annuar, M.N., Pui-San, N.: Working capital management determinants of small and large firms in Malaysia. Int. J. Econ. Manag. **10**(2), 365–377 (2016)

A Study on Strategic Plan for Convergence 6th Industry Policy

Changhwa Baek[1(✉)] and Eunil Son[2]

[1] Department of Industrial Management Engineering, Daejin University, Pocheon, Gyeongi-do, Korea
chbaek@daejin.ac.kr
[2] Gyeongnam 6th Industrialization Support Center, Gyeongnam, Korea

Abstract. This study investigated the concept and characteristics of the 6th industry that is leading new changes in rural areas. The world is rapidly reflecting convergence technology and new trends, and innovative changes are taking place. In rural areas, new innovations are taking place around the 6th industry. Problems with the current policy and support of the 6th industry were examined. In addition, complementary points were derived for these various problems, and measures to overcome them were examined. And through the analysis of relative importance-urgency, an appropriate strategy for problems and supplements was proposed.

Keywords: 6th industry · Policy of the 6th industry · Relative importance-urgency analysis

1 Introduction

Various studies and practices on the 6th industry are being carried out as a way to overcome difficult reality and to give improvement and vitality to rural areas. The 6th industry started in Japan in the early mid-1990s. Rural and fishing villages have developed into a secondary industry that manufactures or processes agricultural and fishing village products based on the primary industry, which is the production activity of existing agricultural and fishing villages. It has also expanded to tertiary industries that provide services such as distribution, sales, and tourism. The 6th industry refers to the activities of agricultural and fishing villages as a comprehensive and multifaceted organism.

The activities of the 6th industry can provide appropriate services or products by understanding and reflecting the rapidly changing trends in rural areas. It can provide new opportunities and values for increasing income and creating jobs for rural communities. Ultimately, it can create a virtuous cycle that leads to the development of farming and fishing villages and further, the growth of the region.

The 6th industry in South Korea conducted research and established strategies centered on the government and local governments. And it is implementing various policies and supports. However, there are many problems that need to be solved from the perspective of people in rural areas and consumers who experience them. Since the role and responsibility of the promotion agency for the 6th industry is not clear, there are

R. Lee (Ed.): ICIS 2021, SCI 1003, pp. 38–49, 2022.
https://doi.org/10.1007/978-3-030-90528-6_4

problems with overlapping related projects. In addition, it is required to implement services that reflect the needs of consumers based on the use of local talents and systematic organization.

The purpose of this study is to examine the government support policies for existing rural and fishing villages and various preceding studies on the 6th industry. And Analyze problems by grasping the current situation. Second, in order to create practical value and welfare in rural areas, we derive implications for the policy of the 6th industry and propose appropriate government policy support.

2 Concept and Characteristics of the 6th Industry

The 6th industry is evolving into services such as various processed goods and sales links, tourism, festivals, etc. by combining with ICT (Information and Communication Technology) through convergence. First of all, it looks at the basic concept of the 6th industry and the characteristics of the 6th industry, which is spreading based on farming and fishing villages.

The 6th industrialization of agriculture is the comprehensive industrialization of agriculture (a comprehensive combination of primary, secondary, and tertiary industries). Agriculture should not only stay in the primary industry, but expand its scope to secondary industries (processing of agricultural and livestock products) and tertiary industries (information service, tourism, etc.). It is a business and activity that brings new values to rural areas and creates new employment opportunities for the elderly and women themselves. The organic combination of the 1st, 2nd, and 3rd industries was identified as 1st industry × 2nd industry × tertiary industry = 6th industry [1].

The term agricultural 6th industrialization was identified as being proposed by Emeritus Professor Naomi Imamura of the University of Tokyo in the early 1990s. Agriculture needs to expand its domain to the secondary industry (processing of agricultural and livestock products) and tertiary industry (agricultural and livestock distribution, tourism, etc.), not just the primary industry (production of agricultural and livestock products). It was summarized as the concept that new value can be created in rural areas and new employment opportunities can be created for women and the elderly only when the comprehensive industrialization of agriculture is implemented [2].

The 6th industry is defined as an industry that increases added value by creating new markets for agriculture and rural areas. This creates new added value by linking agriculture and food, manufacturing and processing of special products (secondary industry) and distribution and sales, culture, experience, tourism, and services (tertiary industry) based on all tangible and intangible resources existing in rural areas. It means an activity to do. The reason why the 6th industry is attracting attention as an alternative to the farmhouse economy is that it is in the process of creating added value that the 6th industry implies. In addition, it is the concept that added value is created when living organisms or resources that become raw materials undergo various treatment processes [3].

When creativity and imagination are added to the agricultural, forestry and livestock products (primary industry), various types of processed products (food, pharmaceuticals, health food, household goods, etc. = secondary industries) are created. It was defined as the development of experiential tourism products (tertiary industry) to circulate and activate it as a marketing strategy. And the 6th industrialization was largely classified into three types [4].

The 6th industrialization was defined as a new value-added activity that processes agricultural products (secondary industry) using tangible and intangible resources existing in rural areas, and links distribution, sales, culture, and experience services (tertiary industry) [5].

It was mentioned that the concept of 6th industrialization is insufficient to explain as a simple set of agricultural production, processing, and service (primary + secondary + tertiary industry = 6th industry). It was explained that it means an organic and comprehensive fusion of these industries (1st × 2nd × 3rd industry = 6th industry). When these three stages have an organic relationship at the local level, their derivative effects can be exerted. It was argued that when agriculture or rural areas (primary industry) decline, the 0th × 2nd × 3rd = 0, and the 6th industry could not be established [6].

The following is a comprehensive summary of prior research and institutional opinions on the concept and characteristics of the 6th industry. First, it is a comprehensive and organic combination of primary industry (rural production), secondary industry (processing and manufacturing), and tertiary industry (services such as distribution, sales, tourism, and experience). Second, it can provide new vitality and opportunities by linking with more systematic and scientific methods based on various tangible and intangible resources and manpower based on rural and fishing villages. Third, through revitalization of the 6th industry, new types of value can be created in farming and fishing villages and new types of job opportunities can be created for the elderly and women. Fourth, by restoring the function of the community in the rural area, it is possible to realize the future welfare of the rural area and establish a foundation for revitalizing the regional economy.

So far, we have looked into the definitions and characteristics related to the 6th industry. Based on this concept, the current state of the 6th industry planned or implemented in rural areas was investigated. In addition, research was conducted on how related policies and supports are being implemented and problems.

3 Review of 6th Industry Prior Research

Among the procedures that are widely used in the field of quality management, the PDCA procedure is used to plan (Plan), execute (Do), review (Check), and then take action (Action). In the 6th industry, it is necessary to carry out various policies and supports using these methods or procedures. Various preceding studies on the 6th industry were conducted to adapt to rapidly evolving technology and diversely changing consumers and environments. We looked at various preceding studies on the 6th industry, which is gradually spreading. In addition, various problems were reviewed by reviewing the current status of support or activities in the 6th industry and related government policies or development policies.

It was pointed out that the existing research focused on the supplier side for quantitative expansion in line with the government policy direction. It emphasized the need to respond to changes in demand because the quantitative growth of rural tourism and the development of high value-added industries are weak, and the foundation for industrialization is also weak [7].

As obstacles to promoting the 6th industrialization, it was found that there are difficulties in smooth financing, solidarity and cooperation among participants in the 6th industry, and difficulties in securing a market for products produced. It was pointed out that as the government's role in revitalizing the 6th industry, there are many demands for expanding policy funding and supporting production channels [8].

It was emphasized that the major problem of the 6th industrialization policy implementation system of Korean agriculture was that the central government was initiated under the overall responsibility with related government agencies. It was pointed out that the establishment of a specific roadmap suitable for regional characteristics and circumstances was insufficient and that various support projects were not being utilized properly [9].

It was pointed out as a problem that the division of roles by actors was not clearly coordinated as other organizations were promoting policy projects. It was pointed out that there was a problem that the nature of the business was duplicated and the efficiency and effectiveness were hindered [10].

Despite the spread of the 6th industrialization, the growth of individual actors tended to stagnate. It was pointed out that the level of 6th industrialization is a supplier-oriented market, which does not respond to diverse and differentiated demands of consumers. He mentioned that it is important to move away from the quantitative expansion strategy of supply-oriented 6th industrialization and promote a consumer-based strategy for related products and services [11].

Since there is a limit to the increase in agricultural income, the importance of diversifying the rural economy was emphasized. He mentioned that it is important to pursue diversification through industrialization of culture and services linked to local agriculture. Through this, he argued that farm household income could also increase and contribute to solving job problems [12].

Through various preceding studies on the 6th industry, the current status and problems were examined. In summary, it was pointed out that there was a low perception of the 6th industrialization and a policy focused on suppliers rather than various demands of consumers. In addition, redundancy and inefficiency were serious due to the lack of support system centered on farming and fishing villages and the promotion of projects with various names for each institution. In addition, the lack of legal equipment and the lack of professional manpower suitable for the 6th industry were pointed out. The lack of cooperation between government departments and the lack of specific strategies tailored to the characteristics of each region were also pointed out as major problems.

Various problems with the policy and support of the 6th industry were pointed out, and various types of prior studies were conducted on surveys and measures to overcome and respond to them. The following is a review and analysis of these prior studies.

As the role of the government to revitalize the 6th industrialization, expansion of policy funds and support for product sales were important, and related coordinators and specialized manpower were emphasized. In addition, he explained that the establishment of an intermediate support organization and expert consulting are important, and it is important to expand educational opportunities for rural industrialization and strengthen support for benchmarking success cases. Therefore, he emphasized the importance of the government's role in disseminating successful cases of 6th industrialization [8].

In order to realize the 6th industry, he stressed the importance of stable income of farmers and consumer-oriented agri-food management. In addition, he emphasized the need to establish comprehensive measures to increase the vitality of rural society and expand the 6th industrialization support system as a future growth engine [13].

Six tasks for revitalizing the 6th industry were presented. First of all, it is important to increase awareness of the 6th industrialization and foster professional manpower, and it is necessary to operate an on-site intermediate support system and related organizations. It also emphasized customized marketing strategies and expanded the 6th industry. In addition, it suggested a plan to establish a network between actors, utilize facilities, and continuously foster primary industries [14].

To investigate the vitalization of the 6th industrialization, in-depth interviews were conducted for workers in the 6th industry, academia, and policy researchers, and several implications were presented through this. The main points emphasized the establishment of smooth communication and the connection between local communities. In addition, It was proposed to establish relationships with consumers and strengthen value-oriented communication as essential requirements [15].

The problem of comprehensive industrialization of agriculture was pointed out and investigated. Through the survey and AHP analysis, the direction of the business promotion of South Korea's 6th industrialization policy was set. In addition, implications for improving the efficiency of implementing related policies were presented. As a result of the final analysis, it was emphasized that the direction of the 6th industrialization policy of agriculture is important to evenly foster the 1st, 2nd and 3rd industries. In addition, he explained that it is necessary to increase the effectiveness of the policy project by focusing more on the secondary processing industry [10].

So far, we have looked at various problems related to the 6th industry policy and support, as well as research on implications and ways to overcome them. In summary, interest and research on the 6th industry policy are receiving great attention, but there are still many points to improve and supplement the reality. Therefore, research on improvement and supplementation of 6th industrial policies and systems for systematic development of agricultural and fishing villages in the future is urgently needed.

4 Analysis and Suggestions

4.1 Problems and Implications of the 6th Industry

Rural and fishing villages have implemented various systems and activities through the introduction and activation of the 6th industry in order to adequately respond to aging and rapid environmental changes. Although it has achieved some positive effects, there are many problems and improvements that need to be supplemented when looking at the related status.

According to the results of the survey on the status of the 6th basic industrialization survey, the rate of satisfaction with policies and difficulties for agricultural and rural businesses was 32.8% of the 'extremely satisfied' and 'satisfied' businesses. As for the difficulties, 'facility funds such as facility expansion' was selected as the top priority in the experience of promoting the 6th industry. In the case of businesses promoting the 6th industry, "facility funding such as facility expansion" was selected, and the most unpromoted businesses appeared in "Rural Education Promotion". It was judged that the 6th industry certified business needs to develop in that it is over 80% in terms of policy satisfaction. As a result of the survey on the satisfaction level of the 6th industry-promoting business policies, the total ratio of 'very satisfied' and 'satisfactory' of the 6th industry certified businesses was 48.9% [3].

The various situations and problems of the 6th industry policy and support were summarized in a comprehensive way. To summarize the main problems, first of all, the development of high-value-added industries was insufficient due to the weak industrialization base, focusing on quantitative growth. In addition, due to the lack of awareness of the 6th industrialization, there were difficulties in related sales channels and marketing strategies, and there was a shortage of professional manpower in rural areas.

In addition, the establishment of a systematic roadmap tailored to the characteristics of each region was insufficient, and differentiated support plans for each region were not implemented. The division of roles for each institution in policy promotion was unclear, so cooperation between ministries and intermediary support organizations were not smoothly implemented and above all, the legal and institutional arrangements were inadequate, and the enactment of the Promotion Act was urgent.

In addition, it was pointed out that it was difficult to secure smooth funding or financial resources related to the 6th industry by promoting the business under various names, Problems arose due to the redundancy and inefficiency of the project. In addition, due to the lack of awareness of the reality of the farming and fishing villages and the lack of a support system centered on the field, the ripple effect within the region was limited and limited. The problem was pointed out as a problem with a supplier-oriented market policy that failed to accommodate the consumer's demands and the limitations that were promoted only with the support of government projects. Various situations and problems related to the 6th industry policy and support have been comprehensively summarized.

Table 1 summarizes the various situations and problems related to the 6th industry policy and support.

The complementary points were studied for the problems to be corrected through the introduction and progress of the 6th industry. There is a need for cooperation that can coexist with each other based on innovative technologies and community solidarity. And it is necessary to review and propose a method that can provide useful benefits to both rural and fishing villages and consumers.

According to the results of the future expansion or reduction of the 6th Basic Survey on Industrialization, 63.4% of businesses promoting the 6th industry in their future business expansion and reduction plans chose business expansion' the most. In addition, 54.2% of businesses that did not implement the 6th industry were "maintained" [3].

When expanding the business, 63.9% of the businesses that promoted the 6th industry chose tertiary industry' the most. 53.8% of businesses that did not promote the 6th industry also showed up in the tertiary industry'. In addition, 68.2% of businesses promoted the 6th industry were reduced to "secondary industry," and 67.9% were reduced to "secondary industry". As for the reason for the reduction, 42.9% of businesses promoting the 6th industry had a "shortage of experts", and 48.4% of businesses that did not promote the 6th industry had a "operation cost burden".

Table 1. Summary of research on problems of 6th industry policy

Researcher	Problems
Park, Mi Og [6]	• Inadequate legal and institutional systems • Lack of monitoring system • Lack of skilled workers in rural areas • Lack of expertise and lack of publicity
Oh, Hyun Sug [12]	• Improvement of co-brand organization development system • Promotion only for government funding • Establish urgent related upbringing law • Town center business center
Jung, Do Chae et al. [11]	• Inadequate response to various needs of consumers • Lack of self-sustenance by many businesses • Growth stagnation of individual subjects • Inadequate field-based support system • Supplier-oriented market policy
Park, Jung Il et al. [10]	• Role sharing by unspecified proponents • Business personality duplication and inefficiency • Promotion of other policy projects by institution
You, Hag Yul [2]	• Management distribution and marketing strategy • Mandatory mid and long term planning • Expanded recognition of 6th industrialization • Lack of interim support organization • Training of professional manpower
Kim, Yong Ryul et al. [9]	• General responsibility of various government agencies • Low utilization of various support projects • Inadequate specific roadmap for regional characteristics • Policy implementation system problem
Jo, Rog Hwan et al. [8]	• Difficulty securing production product line • Expanding government funding • Differentiated support plan by region • Administrative support Intermediate support organization required • Solidarity and Cooperation among Participants
Park, Si Hyun et al. [7]	• Weakened industrialization base • Inadequate development of high-value-added industries • Need to respond to demand changes • Quantitative Growth

For future business expansion and reduction plans, the 6th industry certified businesses most often chose business expansion' at 77.6%, and the 6th industry non-certified businesses chose' business expansion' the most at 59.5%. Businesses that were certified for the 6th industry had an expansion of 75.0% to the "tertiary industry," and those not certified for the 6th industry showed that the secondary industry and the tertiary industry were 50.0%, respectively. In addition, 68.2% of the business areas to be reduced for non-certified businesses in the 6th industry were secondary industry' and 31.8% of tertiary industry'. For the reasons of the reduction, 46.2% of businesses that were not certified in the 6th industry had a "shortage of experts", 30.8% of the "operating expenses", and 23.1% of the "capacity limit".

Table 2 summarizes the implications for the 6th industry policy and support and future supplements. It is important to increase awareness of the 6th industry to form value relationships with consumers and to stabilize the income of rural communities and increase vitality. In addition, by fostering coordinators and experts related to the 6th industry, they utilize the spread of success cases. Therefore, it is necessary to establish specialized programs and management systems that expand related educational opportunities. In addition, it is necessary to expand the area of the 6th industry using culture and establish a long-term vision. Consumer-oriented agri-food management, customized marketing strategies, and expansion of sales networks are needed. It is also important to quickly reorganize related laws and systems and establish an on-site intermediary support system to designate and operate an overall organization. In addition, it is necessary to select the projects to focus on by sector, along with the even development of the primary, secondary, and tertiary industries. It is important to revitalize the support for diversification of the rural economy by expanding support for related government policy funds. It is also necessary to manage partnerships with specialized organizations by establishing linkages between local communities and establishing an aggregate between local governments, local residents, and business entities.

In addition, it is necessary to maintain energy self-sufficiency and eco-friendly rural spaces in rural areas. Based on this, it is important to diversify the regional industrial structure through convergence in various fields.

Ultimately, it is very important to supplement the system of the 6th industry to create jobs for the elderly or women, and to restore local communities by improving income. It is very necessary to create new value through the development and branding of new products and new market development.

4.2 Policy Support Strategy for the 6th Industry

Based on the problems, implications, and plans for the 6th industry's policies and support, an analysis of the relative importance and urgency of each item was conducted. The contents of this are illustrated in Fig. 1. This allows potential factors for each item to be evaluated on a relative basis to prioritize items and derive strategies for each item-specific field.

First of all, legal and institutional arrangements are needed in areas of relatively high importance and high urgency. In addition, the enactment of related development laws is urgent, and expansion of government funding is important.In addition, a clear division of roles for each subject and an on-site intermediate support system are required, and it

Table 2. Summary of study on complementary points of 6th industrial policy

Researcher	Complementary points
Park, Mi Og [6]	• Strengthening human resource development programs • Building programs and management system • Management by multilateral partnership of specialized agencies • Linking Support Subjects
Park, Jung Il et al. [10]	• Increase efficiency of policy business by focusing more on secondary processing industry • Properly cultivating the first, second, and third industries • Selection of priority business by sector
Jung, Oh Rag et al. [16]	• Establishment of governance among local governments - local residents - business entities • Professional role and system construction such as coordinator • Institutional supplementation • Diversification of regional industrial structure through confluence
Oh, Hyun Sug [12]	• Construct circulation economy • Expansion of near-field sales network • Active support for rural economic diversification • Improvement of eco-friendly rural space
Yoo, S. U. and Roo, S. Y. [15]	• Strengthening value-oriented communication • Enhancing the relationship between stakeholders • Strong infrastructure and expertise • Form relationship with consumers • Establish long-term vision
Gwon, Yong Dug [17]	• Establish a general organization to perform interim support functions • Strengthening and cooperating with local sovereignty • Improvement of related laws and systems • Establishment of middle core organization
Yun, H and Lee, Y. O. [14]	• Expansion of 6th industrial area including agricultural products and culture • Proper organization setup and operation • Expanding awareness of 6th industrialization • Customized distribution and marketing strategies • Training of professional manpower
Jo, Rog Hwan et al. [8]	• Installation of intermediate support organization • Expand educational opportunities and spread success stories • Sales support for produced products • Training coordinator and professional manpower • Expansion of government policy funding

is important to support production, sales, and marketing strategies by fostering experts in the 6th industry. In addition, it is necessary to stabilize the management income of rural areas based on differentiated support measures for each region. There is a need for "an urgent and important strategy" in this area.

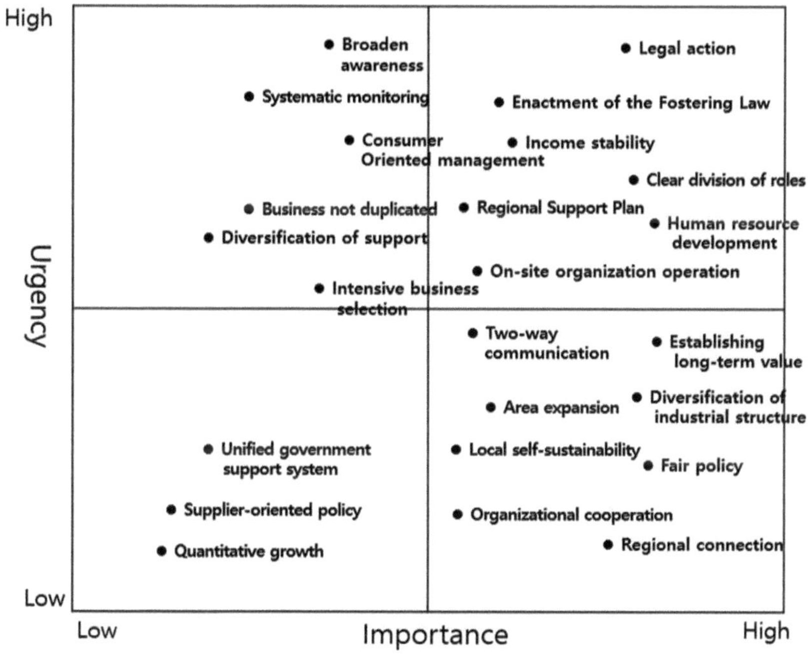

Fig. 1. Strategies of the 6th industry through the importance-urgency analysis

The next area is of high relative importance but low urgency. Based on the establishment of a long-term vision, it is necessary to evenly foster primary, secondary, and tertiary industries and to expand the area of 6th industrialization. In addition, it is necessary to expand the linkage between the self-sustainability of businesses and the local community, and it is important to establish diversification of the local industrial structure through convergence based on cooperation between participants. In addition, it is necessary to reinforce value-oriented communication and improve the awareness and space of eco-friendly farming and fishing villages.

The next area is of low importance but high urgency. It is necessary to develop and manage agri-food to increase awareness of the 6th industrialization and to target consumers. In addition, it is important to eliminate inefficiencies caused by overlapping projects by policy or support, and to select projects to focus on by sector. In addition, it is necessary to activate multilateral support and to establish a monitoring system to properly manage and evaluate it.

Lastly, the next areas are those with low relative importance and low urgency. It pursues quantitative growth and is mainly driven by a supplier-oriented market policy.

In addition, the support system centered on government agencies rather than consumers or on-site should be avoided and improved in the future.

5 Conclusion

Based on the concept and characteristics of the 6th industry that will lead to new vitality and substantial change in the rural area, the problems of the 6th industry's policy and support were examined. And it drew complementary points for these various problems and looked into ways to overcome them. This was suggested a suitable strategy by using the relative importance-urgency analysis.

To summarize the main contents of this, first, a "N(Now)" strategy is needed as an area where both relative importance and urgency are high. The legal and institutional system, the enactment of related fostering laws, and an on-site intermediary support system should be implemented first.

Second, a "T(Timing)" strategy is needed where the relative importance is high in the domain but the urgency is low. In this regard, it is necessary to evenly foster primary, secondary and tertiary industries and to expand the area of 6th industrialization.

Third, a "C (Cooperating)" strategy is needed in one area with low relative importance but high urgency. It is necessary to expand the awareness of the 6th industrialization and develop a business aimed at consumers.

Finally, a "L(Late)" strategy is needed in one specific area where both relative importance and urgency are low. There is an urgent need to pursue quantitative growth or to change supplier-oriented market policies.

As for the limitations of this study and future research tasks, it is necessary to first verify the strategy and plan of the 6th industry policy and support in the future, by a group of experts or related organizations and officials. In addition, it is necessary to select items with high priority in areas of high relative importance and urgency. Through this, further research on specific plans and detailed procedures is needed so that future policies and support for the 6th industry can be practically implemented. In addition, for the practical development and vitalization of the farming and fishing villages, it is necessary to improve the policies and support plans of the 6th industry. Continuous monitoring activities are required based on structuring by item and subject.

References

1. Lee, B.O.: Japan's 6th industrialization development direction and challenges. Korea Rural Econ. Ins. Other Res. Rep. **143**, 211–232 (2012)
2. You, H.Y.: Challenges for necessity and activation of sixth industrialization in agriculture. KRIHS Monthly Magazine 29–35 (2013)
3. Gyeongsangnam-do. Survey on the Basic Status of Sixth Industrialization in Gyeongsangnam-do, Gyeongnam 6th Industry Support Center (2017)
4. Kim, T.G.: 6th industrialization concept and establishment method of agriculture. Korea Rural Econ. Ins. (2013)
5. Ministry of Agriculture, Food and Rural Affairs.: Creation in agriculture and rural areas: Best Practices for 6th Industrialization (2014)

6. Park, M.O.: Green revolution in the 21st century, case study on 6th industrialization policy. Korean Soc. Admin. Res. **28**(2), 111–144 (2017)
7. Park, S.H.: New directions and policy tasks for rural tourism. Korea Rural Econ. Ins. Basic Res. Rep. 1–221 (2012)
8. Jo, R.H.: Analysis of the sixth industry policy needs in agriculture and rural areas. Rural Tour. Res. **20**(2), 137–152 (2013)
9. Kim, Y.R.: 6th Industrialization and policy tasks of Korean agriculture. Korea Rural Econ. Ins. (2014)
10. Park, J.I.: Prioritization of the sixth industrialization policy of agriculture: using AHP. Agric. Sci. Res. **42**(4), 461–467 (2015)
11. Jung, D.C.: Direction and tasks of 6th industrialization in regional unit. Korea Rural Econ. Ins. 1–24 (2016)
12. Oh, H.S.: Rural development policy to create added value. Insight GSnJ. 1–16 (2017)
13. Kim, S.H.: Survey Results and Improvement Plan for Sixth Industrialization Regulation. Ministry Agric. Livestock (2014)
14. Yun, H.Y., Lee, Y.O.: 6th industrialization for the realization of creative agriculture. Korea Rural Econ. Ins. Other Res. Rep. 703–730 (2014)
15. Yoo, S.U., Roo, S.Y.: Promotion of agricultural value and PR communication through activation of 6th industry. Korea Ins. Public Inf. (2015)
16. Jung, O.R.: A study on the activation plan of local small and medium sized city utilizing 6th industrialization. Resident. Environ. **13**(4), 255–266 (2015)
17. Gwon, Y.D.: 6th industrialization of agriculture and its tasks. Gyeongnam Develop. 57–70 (2014)

Development of Air Pressure Measurement System of Suction Cups in a Vacuum Gripper

Sujeong Baek[1(✉)], Dong Oh Kim[1], Seo Jin Lee[1], Na Hyeon Yu[1], and Su In Chea[2]

[1] Department of Industrial and Management Engineering, Hanbat National University, Dongseodae-ro 125, Yuseong-gu, Daejeon, Republic of Korea
sbaek@hanbat.ac.kr, {20151470,20174038, 20184039}@edu.hanbat.ac.kr

[2] GL Company, 303, Daedong-ro, Sasang-gu, Busan, Republic of Korea
gl@glcomany.kr

Abstract. A vacuum gripper is typically used in various industrial fields for tasks such as clamping and lifting. This type of gripper works with suction cups, which are connected through a vacuum line to pick up a target part. During the operation, vacuum grippers sometimes fail to pick up a part properly because of defective surface conditions or unoptimized control variables for different products. However, it is not easy to know whether the current operation will be correctly performed via a traditional gripper. Therefore, we aim to identify relevant parameters for the real-time condition of a vacuum gripper system to detect the corresponding operation status. An air pressure measurement is developed to monitor the degree of change in air pressure in order to determine whether a target product can be held by a vacuum gripper. A robotic arm with one suction cup is utilized, and an analog air pressure sensor is installed for real-time condition monitoring of the gripper operation in a vacuum line. Using this measurement system, we can conduct an experiment to discover effects of control parameters to pick-up operation of a vacuum gripper, whether a gripper succeeds or fails in holding a part.

Keywords: Air pressure measurement · Condition monitoring · Suction cup · Vacuum gripper

1 Introduction

Mass customization and personalization has become a trend of current manufacturing, with various types of products simultaneously produced in the same production line [8, 10]. A vacuum gripper is popularly used as the end-effector of several actuators to effectively handle different types of parts or products [2, 16, 20] such as semi-conductor wafers [21], cartons [18], fabrics [17], or foods [5, 6]. A vacuum gripper has many advantages over mechanical types of grippers, which mimic human fingers and are usually employed to pick up a fixed size and shape product [15, 22]. These include simple installation and operation and easy application for different shapes and hardness of products [3, 4]. Vacuum grippers work with more than one suction cup, which are

R. Lee (Ed.): ICIS 2021, SCI 1003, pp. 50–61, 2022.
https://doi.org/10.1007/978-3-030-90528-6_5

connected through a corresponding vacuum line to an air compressor, in order to enable lifting a target part using near zero vacuum pressure [19]. The vacuum is generated using the ejector principle [7, 12] and the detailed procedure for picking up a part using a vacuum gripper is described below.

A Venturi channel is originally devised for increasing the speed of the inlet airflow and looks like two cones facing each other with a narrow channel connecting them. Therefore, from the view of inlet airflow, the cross-section of the pipe becomes smaller and then suddenly larger [7]. The first cone is located in the front and its smaller, and this cross-sectional area is called a jet nozzle. The second cone section is located at the rear and its larger, and this cross-sectional diameter is called the receiver nozzle. When the compressed air travels from the vacuum generator through the jet nozzle of a Venturi channel to the receiver nozzle, the speed of inlet airflow will increase. Consequently, the pressure at the downstream of the jet nozzle will become lower than the atmospheric pressure. At this step, air is sucked from the additional narrow air line that connects to each suction cup, causing a partial vacuum in the line.

Although vacuum grippers are simple to install, use, and maintain, they sometimes fail to pick up a part correctly due to the surface condition of the part or unoptimized control variables for different types of transported parts. In the case of the surface condition of the target parts, there has been much research focused on detecting the condition before the operation process [1, 14]. For example, Jeon and Song (2019) developed an artificial neural network model in order to classify a box cover as normal or defective [13]. They measured the relative distance between the sensor itself and the top surface of the box cover and used the dataset as input for the developed model. As a result, a defective box cover that could not be picked up by a vacuum gripper could be removed before the operation process. A tiny particle on the target surface of the product can easily generate an error situation during an operation process. Therefore, Fisher et al. (2019) analyzed the impact of a vacuum gripper used for wafer handling by varying the particles on the wafer to control surface cleanliness (*e.g.*, contaminated or cleaned gripper surface) [9]. They found that the number of particles and particle size were significantly related to the performance of suction cups.

For the machine control parameters of a vacuum gripper, cup size and the number of suction cups determines whether the vacuum gripper can or cannot lift the target object. For example, the size of vacuum cups is highly related to the power of suction [11]. A larger cup is used to provide a larger amount of suction power. However, the cup size is related to the performance of the connected air compressor and its change could result in high installation and operation costs. Alternatively, if the cups are too small for a vacuum gripper, the operation process cannot satisfactorily be performed. These types of machine control parameters do not change during the operation process because their control values are optimized and fixed to the target products before the operation process starts [23]. Unfortunately, the gripper still sometimes fails to handle some products with the predefined parameter values and it is not simple to identify the suction status of the vacuum gripper before the pick-up process starts. Consequently, this results in sudden or unexpected stops in the manufacturing process which are costly and inefficient.

Therefore, it is necessary to identify appropriate machine control parameter and sensor signals which can be monitored in real-time and changed depending on the current operation status. We developed an air pressure control and measurement system to monitor machine control parameters immediately (in particular, the degree of change in air pressure). To accomplish this, a new air pressure sensor and vacuum switch are added to monitor the current vacuum gripper working status in real-time. A simple 4-axis robotic arm with one suction cup was used to demonstrate the developed system by picking up a box, therefore we can also easily control the position parameters of a suction cup in the vacuum gripper. In addition, we conduct an experiment to find statistical effects of control parameters to the performance of pick-up operation by vacuum gripper.

The remainder of this paper is organized as follows: Sect. 2 describes the developed system configuration for condition monitoring of the machine control parameters of a vacuum gripper. Section 3 introduces the data acquisition software to monitor the pressure signals of the airflow in a Venturi channel during one operation process cycle. Section 4 describes our experiment to determine if the effects of control parameters (such as an amount of initial pressure, flatness of a contact surface, suction cup size, and location(s) of suction) are significant on the normal pick-up process operation Finally, in Sect. 5, the conclusion and further research opportunities are presented.

2 Hardware Configuration for Condition Monitoring of Vacuum Gripper

We first determined an appropriate machine control parameter that could easily be changed between operation processes. For example, several cup parameters were excluded as target machine parameters in this study, because it is difficult to change the cups between operation processes. As a result, we selected the amount of inlet air pressure, to the suction cup, as our machine control parameter. Consequently, air flow was chosen as the machine monitoring parameter for the operation process of picking up a part, and the related sensor signal in real-time was monitor, which will be described in Sect. 3. The system developed for the condition monitoring of a vacuum gripper is shown in Fig. 1 and Fig. 2.

A conventional air service unit and compressor are installed for the normal operation of a vacuum gripper. A four axes robotic arm (Dobot Magician) is used as the main actuator to perform the pick-up process. A vacuum gripper with one suction cup is applied as its end effecter. A power supply and converter are also equipped to move the actuator and collect the relevant sensor signals from the newly equipped analog pressure sensor (Autonics' PSQ-C1CU-R1/8) in the vacuum line. The controller (Arduino Uno with ethernet commutation module), illustrated in Fig. 2(a), is only used for collecting the relevant airflow signals, not for controlling the robot which is controlled by the connected PC-based controller. A vision camera is installed to operate the robot more intelligently. 2D-image data from the vision camera is used to automatically analyze the center of mass of the target product in the working spaces. Subsequently, the robot will move to the corresponding x, y location and conduct the pick-up process. In addition, there are other related components which are commonly used for a vacuum gripper, such as a single solenoid valve (directional control valve) and a vacuum switch (Shinshin electrical company's SV-15HS).

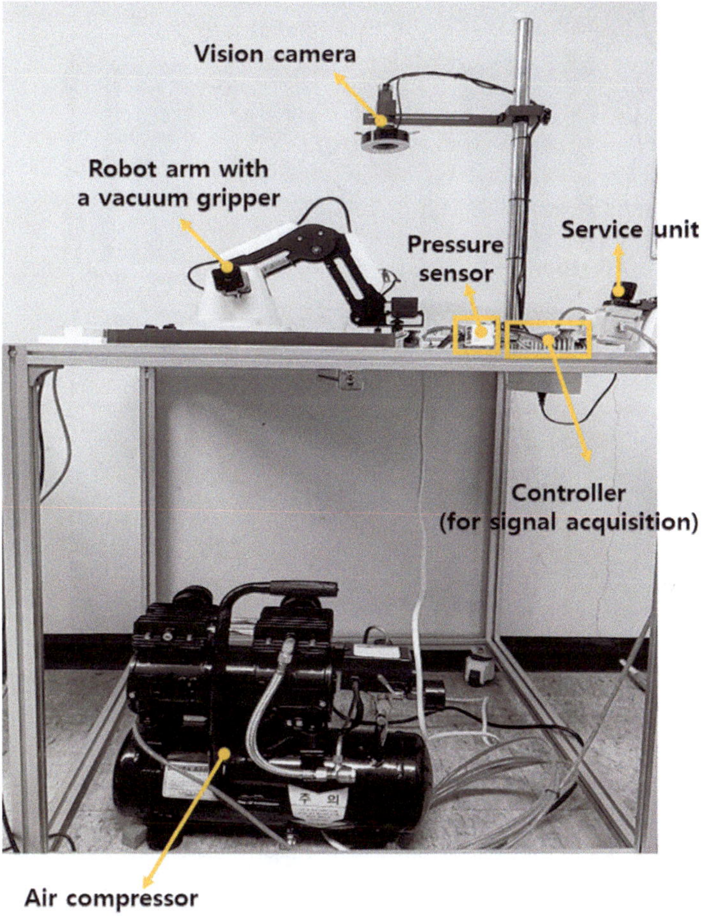

Fig. 1. The entire (front) view of the proposed air pressure measurement system

3 The Developed Air Pressure Signal Acquisition System

The proposed air pressure control and measurement system is used to identify whether the vacuum gripper can pick up a part before lifting, as a condition monitoring module of a vacuum gripper. First, every component is turned on including the robot, vision camera, connected PC-based controller, and air compressor. The developed software, as illustrated in Fig. 3(a), is used to connect the data acquisition controller.

The PC-based controller is easily connected to the data acquisition controller by inputting the correct IP address of the data acquisition controller and ethernet port. Subsequently, when a user commands the robot to pick up a part in the working space, as depicted in Fig. 4(a), the robot and vision camera receive the command and start the operation process. For example, in Fig. 4, a red colored box is the target part to be picked up by the robot. The vision camera provides the target location to the robot by analyzing

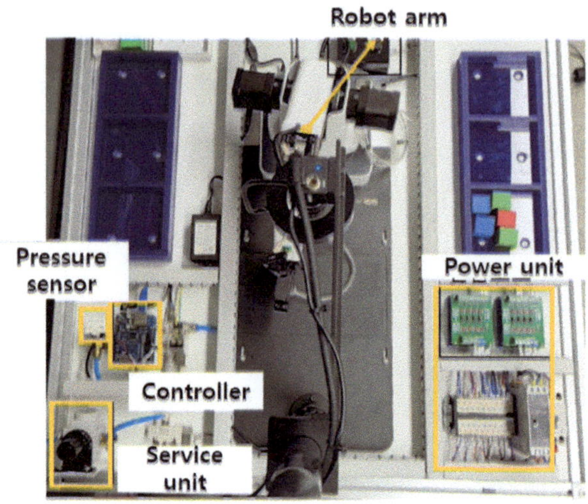

(a)

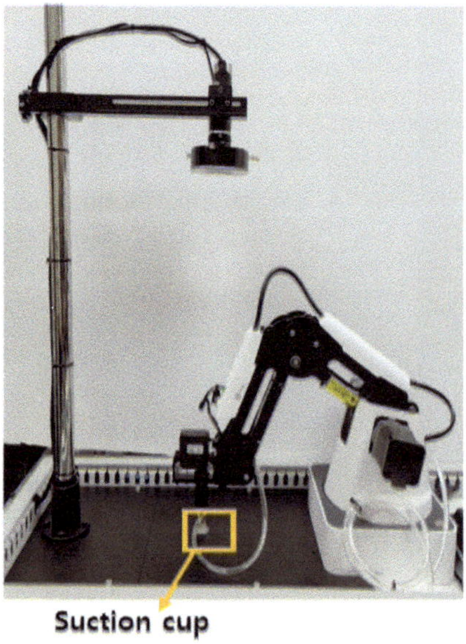

(b)

Fig. 2. The detail view of the proposed air pressure measurement system: (a) top view with each component description and (b) robot arm with a single suction cup which can conduct pick up operation automatically

current 2D images and calculating the center point of the part. Then, the robot arm will move to the corresponding location and downward to pick up the part.

When the suction cup contacts the target surface of the part, the vacuum generator starts to operate and our analog sensor signal for the pressure of the connected Venturi

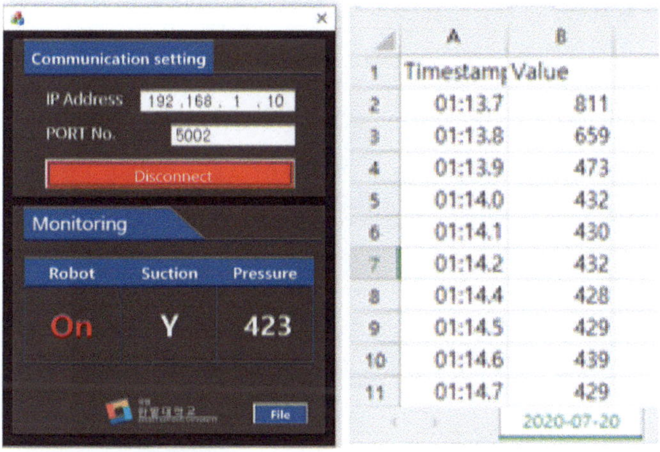

	A	B
1	Timestamp	Value
2	01:13.7	811
3	01:13.8	659
4	01:13.9	473
5	01:14.0	432
6	01:14.1	430
7	01:14.2	432
8	01:14.4	428
9	01:14.5	429
10	01:14.6	439
11	01:14.7	429

2020-07-20

(a) (b)

Fig. 3. Corresponding control software and measurement results. (a) the developed control software to collect analog airflow and digital vacuum switch signals in real-time, and (b) the collected analog airflow sensor signal over a time span

channel is collected automatically and in real-time. The real-time measurement values are also monitored in the developed software, as shown in Fig. 3(a). When the suction operation of the vacuum gripper succeeds, the robot will move and put the part in the pre-defined magazine, as illustrated in Fig. 4(b) and (c). After finishing each cycle of the operation process, the robot will move to the initial position for the next command and the collected analog sensor signals are exported to the Excel spreadsheet to save as history data with a corresponding time stamp, as shown in Fig. 3(b). The duration time for picking up and moving the target part is approximately 5 s and the analog signal is recorded at close to an about 10 Hz sampling rate.

Figure 5 shows the corresponding signals of air pressure to the suction cup during one operation process cycle (illustrated in Fig. 4). The initial pressure of inlet airflow in a Venturi channel is almost identical to the originally provided air pressure supplied by the air service unit and air compressor, before the vacuum gripper starts to suction. When the controller starts to lift part, by generating partial vacuum in the gripper, the monitored air pressure continues to decrease until a certain level, between a point (a) and (b) in Fig. 5. This operation usually takes about less than a second. Then, the decreased air pressure is constantly kept during lifting and moving of the target to the predefined destination. Although there are some fluctuations in the inlet pressure during the lifting operation, as indicated from time points (b) to (c) in Fig. 4, it does not indicate failure of the pick-up process and we can consider that the lower pressure is properly maintained. At the time stamp (c), the partial vacuum disappears to put the part on the workplace and signal recording is also completed, because the actuator finishes to move to the destination, as described in Fig. 4(c).

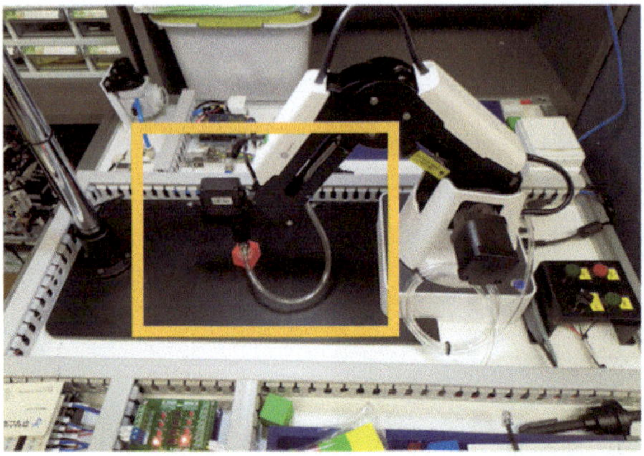

(a)

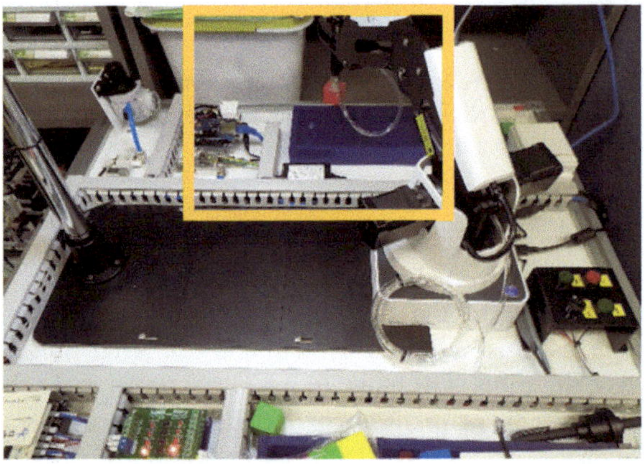

(b)

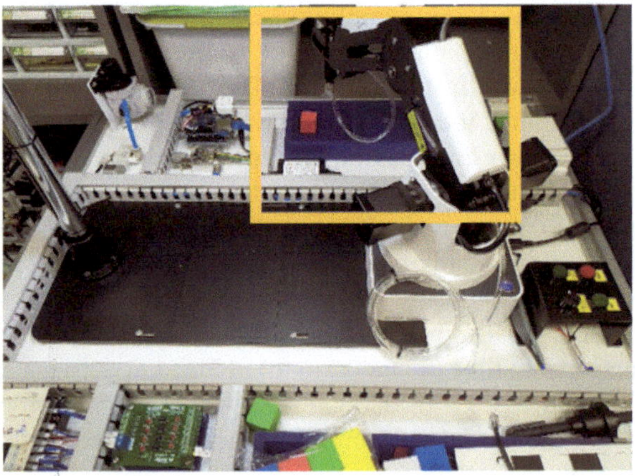

(c)

Fig. 4. Description for how to conduct the given a pick-up operation process. (a) the robot moves to target position and starts to suction, (b) the robot moves to the destination position, and (c) target is correctly put in the destination position.

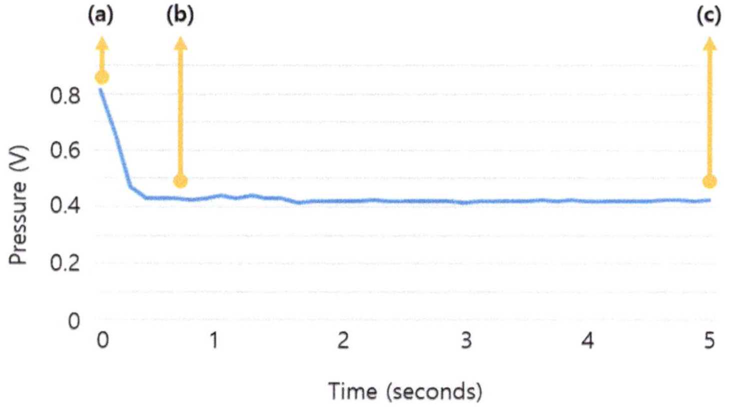

Fig. 5. Corresponding signal trends of the airflow pressure during one cycle of the pick-up operation process. (a), (b), and (c) is time stamp mark that indicates the time when the corresponding step of the operation process is performed in Fig. 4(a), (b), and (c), respectively

4 Process Operation Deepening on Initial Pressure Value

We conducted an experiment to determine if the effects of the initial pressure of the Venturi channel are significant on the normal pick-up process operation. To conduct the experiment and compare the result, we used another vacuum gripper with 4 suction cups, as shown in Fig. 6. The identical box pick-up operation was performed by the two different types of vacuum grippers and the detailed information is summarized in Table 1.

In addition, two boxes were used as the target parts, the first box (called as Box 1) is always considered as a normal quality to pick up by the vacuum gripper in the industrial robot (which is the current operation condition), which has sufficiently flat top surface. The second box (called as Box 2) are also treated as normal quality in product quality inspection stage (before being put into the pick-up process), but it usually results in pick-up failure situation because the type of box has not flat enough to use the vacuum gripper. Therefore, although Box 2 is not considered as defect, but causes hidden fault in the manufacturing process.

Experimental results are summarized in Table 2. We conducted several replications for identical set-up, but there is no with-in group difference. In the case of the proposed actuator (relatively larger suction cup), there was no variation depending on the initial air pressure and the degree of flatness of box contact surface. They could pick-up any transported part with the little amount of inlet air pressure. However, in the case of the vacuum gripper in the industrial robot, Box 1 was barely able to be picked up when less than 0.4 MPa is provided. Even, in the case of Box 2, the vacuum gripper of the industrial robot has never been successfully picked up. From this result, we identified that supplied initial pressure to suction cups did not significantly affect the conduction of the pick-up operation if the target surface was quite flat for contact with the suction cups. In addition, it was observed that the performance of the pick-up operation by a vacuum

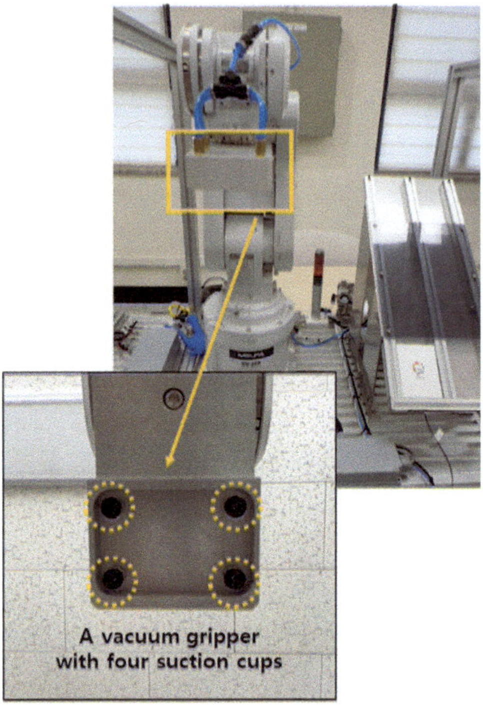

Fig. 6. Industrial robot (manufactured by Mitsubishi Electric) with vacuum gripper with four suction cups

Table 1. Specification of two vacuum grippers used in the experiment

	The proposed system	Industrial robot
The number of suction cups	1 cup	4 cups with a shape of 2 by 2 array
Diameter of each suction cup	2 cm	1 cm
The contact position for suction on a target product	At the center of mass	Near the four corners

gripper could be improved by adjusting relevant machine parameter of the gripper, once a contact surface of a product passed the quality inspection.

Table 2. Specification of two vacuum grippers used in the experiment

Initial pressure (MPa)	The proposed system		Industrial robot	
	Box 1	Box 2	Box 1	Box 2
0.1	1	1	0	0
0.2	1	1	0	0
0.3	1	1	0	0
0.4	1	1	1	0
0.5	1	1	1	0
0.6	1	1	1	0

*0 and 1 indicate failure and success in picking up a part, respectively.

5 Conclusion

We developed an air pressure control and measurement system for conducting condition monitoring of a vacuum gripper. To demonstrate the air pressure control and measurement system, we used a simple 4-axis robotic arm with one suction cup, installed with normal corresponding additional equipment such as a power supply unit and service module. A vacuum generator creates a vacuum condition in the corresponding air flow line when the robotic arm starts to pick-up and move a target product to the destination. The developed system automatically recorded the analog measurements of air pressure signals. Finally, analog signals were monitored in the developed software in real-time, collected by a controller, and exported to a spreadsheet through ethernet communication. In addition, it was observed, by analyzing historical signal data from the developed system, that the size of suction cup and location(s) of suction have significant effects to generate a sufficient vacuum condition in vacuum gripper for lifting operation, rather than an amount of input pressure and flatness of a contact surface.

Further studies need to be conducted to improve the function of the proposed system, such as: i) detecting suction status earlier to provide more preparation time to workers, ii) generalizing the results for different shapes, sizes, and weights of target parts, and iii) upgrading the system for a vacuum gripper with different cup specifications, such as size and number.

Acknowledgments. This work was supported by the National Research Foundation of Korea (NRF) grant funded by the Korea government (MSIT) (No. NRF-2019R1G1A1097478), and was also supported by Korea Institute for Advancement of Technology (KIAT) grant funded by the Korea Government (MOTIE) (P0012744, The Competency Development Program for Industry Specialist).

References

1. Baek, S., Jeon, S.H., Song, E.C.: Real-time monitoring of a vacuum gripper pick-up operation in a box packaging process by CNN based the box surface deviation image analysis. J. Korean Inst. Ind. Eng. **46**(2), 107–113 (2020)
2. Bi, Z.M., Zhang, W.J.: Flexible fixture design and automation: review, issues and future directions. Int. J. Prod. Res. **39**(13), 2867–2894 (2001)
3. Callies, R., Fronz, S.: Recursive modeling and control of multi-link manipulators with vacuum cup grippers. Math. Comput. Simul. **79**, 906–916 (2008)
4. Cho, Y., Lee, H., Gong, J.: Development of the flexible gripping system for automotive battery. In: Proceedings of 2018 Fall Conference of the Korean Society of Automotive Engineers, pp. 1052–1053 (2018)
5. Chua, P., Ilschner, T., Caldwell, D.G.: Robotic manipulation of food products - a review. Ind. Robot **30**(4), 345–354 (2003)
6. Davis, S., Gray, J.O., Caldwell, D.G.: An end-effector based on the bernoulli principle for handling sliced fruit and vegetables. Robot. Comput. Integr. Manuf. **24**(2), 249–257 (2008)
7. Dini, G., Gantoni, G., Failli, F.: Grasping leather plies by bernoulli grippers. CIPR Ann. **58**(1), 21–24 (2009)
8. Dufe, N., Bendul, J., Knollmann, M.: An analytical approach to improving due-date and lead-time dynamics in production systems. J. Manuf. Syst. **45**, 273–285 (2017)
9. Fischer, A., Moldovan, A., Rentsch, J.: Impact of vacuum grippers utilized for automated wafer handling prior a-Si passivation for silicon heterojunction solar cells. In: Proceedings of AIP Conference, vol. 2147, p. 05002:1–6 (2019)
10. Huang, S., Wang, G., Shang, X., Yan, Y.: Reconfiguration point decision method based on dynamic complexity for reconfigurable manufacturing system (RMS). J. Intell. Manuf. **29**(5), 1031–1043 (2017). https://doi.org/10.1007/s10845-017-1318-2
11. Jaiswal, A.K., Kumar, B.: Design construction of vacuum cup gripper of robots - as a pick and place operation tool. Int. J. Sci. Technol. **6**(4), 26–35 (2016)
12. Jaiswal, A.K., Kumar, B.: Vacuum cup grippers for material handling in industry. Int. J. Innov. Sci. Eng. Technol. **4**(6), 187–194 (2017)
13. Jeon, S.H., Song, E.C.: Development of real-time and automated quality inspection process using machine learning technique. In: Proceedings of 2019 Fall Conference of Korean Institute of Industrial Engineers, pp. 3720–3734 (2019)
14. Jeon, S.H., Song, E.C., Baek, S.: Implementation of automated quality monitoring process for a defective box by analyzing ultrasonic sensor signals. In: Proceedings of 2019 Summer Conference of Society for Computational Design and Engineering, p. 196 (2019)
15. Lee, J.K., Kim, H.J., Lee, S., Lee, K.C.: Design of grasping objects system of robot gripper using artificial neural network. In: Proceedings of the Korean Society of Manufacturing Process Engineers Autumn Conference, p. 109 (2014)
16. Manitriota, G., Messina, A.: Theoretical and experimental study of the performance of flat suction cups in the presence of tangential loads. Mech. Mach. Theory **46**(5), 607–617 (2011)
17. Ozcelik, B., Erzincanli, F.: Examination of the movement of a woven fabric in the horizontal direction using a non-contact end-effector. Int. J. Adv. Manuf. Technol. **25**(5–6), 527–532 (2005). https://doi.org/10.1007/s00170-004-2075-x
18. Ozcelik, B., Erzincanli, F., Findik, F.: Evaluation of handling results of various materials using a non-contact end-effector. Ind. Robot **30**(4), 363–369 (2003)
19. Rateni, G., Cianchetti, M., Ciuti, G., Menciassi, A., Laschi, C.: Design and development of a soft robotic gripper for manipulation in minimally invasive surgery: a proof of concept. Meccanica **50**(11), 2855–2863 (2015). https://doi.org/10.1007/s11012-015-0261-6

20. Schaffrath, R., Jäger, E., Winkler, G., Doant, J., Todtermuschke, M.: Vacuum gripper without central compressed air supply. Procedia CIRP **97**, 76–81 (2021)
21. Schmalz, K.: Vacuum grippers on robots and handling systems. In: Proceedings of the 25th International Symposium on Industrial Robots, pp. 59–65 (1994)
22. Seo, J.W., Lee, J.K., Lee, S., Lee, K.C.: Slip detection of robot gripper with flexible tactile sensor. J. Korean Soc. Precis. Eng. **31**(2), 157–164 (2014)
23. Valencia, A.J., Idrovo, R.M., Sappa, A.D., Guingla, D.P., Ochoa, D.: A 3D vision based approach for optimal grasp of vacuum grippers. In: Proceedings of 2017 IEEE International Workshop of Electronics, Control, Measurement, Signals and Their Application to Mechatronics, pp. 1–6 (2017)

Efficiency Assessment in Digital and Online Functions of University Libraries Using Data Envelopment Analysis

Youn Sung Kim[1], Seungbeom Kim[2(✉)], Kyung Mi Bae[3], and Min Seo Park[1]

[1] College of Business Administration, Inha University, Incheon, Korea
motbeol@inha.ac.kr, 22201580@inha.edu
[2] College of Business Administration, Hongik University, Seoul, Korea
sbkim@hongik.ac.kr
[3] Department of International Business and Trade, School of Global Convergence Studies, Inha University, Incheon, Korea
gracebae@inha.ac.kr

Abstract. Libraries are encountering certain challenges derived from changes on their business environment. With the advent of COVID-19 pandemic, digital and online functions of university libraries in Korea is suggested as an alternative for the future. Thus, it is significant to assess libraries' business efficiency focusing on digital function while it is able to acquire valuable factors for future strategy. Therefore, the study adopts DEA (Data Envelopment Analysis), a widely accepted analysis model for measuring efficiency. Moreover, the study derived appropriate factors for efficiency assessment in digital service of university library by literature review. Through this, factors regarding efficiency in university libraries can be measured. We hope this study could contribute to identify inefficient factors in university libraries as well as gaining basic data for developing improvement strategy for future.

Keywords: Efficiency assessment · Data Envelopment Analysis · University library · Digital library · Online library · E-book

1 Introduction

Management techniques are increasingly recognized as important in the organizational management of libraries. In particular, it can be said that university libraries are a service provider to meet customer needs [1]. In accordance with this stance, a number of preceding studies have worked on management techniques such as CRM [2], SERVQUAL [3], and other management techniques in university libraries. In addition, with the development of the service industry, the service innovation of university libraries was also attracted attention. Studies were conducted on the role of knowledge management in university library service innovation [4], and a framework for value co-creation for university library service innovation [5].

This research is supported by Inha University.

R. Lee (Ed.): ICIS 2021, SCI 1003, pp. 62–74, 2022.
https://doi.org/10.1007/978-3-030-90528-6_6

As the management analysis of university libraries was conducted, interest in the efficiency of university library organization also increased. Efficiency evaluation was mainly a major concern for for-profit organizations, but now it is also an important issue for non-profit organizations as well. Accordingly, interest in the methodology for measuring efficiency of non-profit organizations has also increased [6, 7].

The Data Envelopment Analysis (DEA) model is mainly used to measure the efficiency of libraries including universities. This is because DEA can provide quantitative data on appropriate and inefficient factors in measuring and evaluating the effectiveness of organizations with multiple input and output factors, such as university libraries [1]. In addition, it is possible to establish future operational strategies by grasping the profitability position based on the results of measuring the efficiency of service companies through appropriate techniques such as DEA [8].

It is important to measure the effectiveness of university libraries because such organizations are in a changing business environment. First, many libraries have been temporarily closed due to the aftermath of the '2019 novel coronavirus (COVID-19)', and changes such as non-face-to-face conversion of service provision for re-opening are indispensable [9]. In particular, in the case of Korean libraries, are in an unstable business environment due to changes in the supply environment, such as a decrease in the amount of reading and a change in preference for reading media, changes in the demand environment and the information media due to technological advancement, intensifying competition among providers, etc. [10].

In these environment dynamics, the digital and online service of library are suggested as a breakthrough. For instance, various libraries have accelerated the digital factor in their services by expanding their digitalized sources and collections of information [9]. In Korea, certain universities are operating a new type of non-face-to-face service that provides simple book reservation and book delivery service for loading and borrowing on their websites [11, 12]. On the other hand, Marker Space is one another technology-based library innovation for university libraries in Korea, while it could be function as a brand new service space with advanced ICT such as 3D printers and virtual reality (VR) [13].

Therefore, it is necessary to measure the operational efficiency especially regarding online and digital services provided by libraries to identify inefficient factors considering the current situation of Korean university libraries. It is important to provide materials that can contribute to the establishment of future operational strategies, such as providing opportunities to benchmark the efficient parts of other libraries. Focusing on the aspect, this study uses the DEA technique for a comprehensive analysis of the efficiency of university libraries and suggests ways to improve efficiency and productivity through it.

In this study, firstly, the concept related to DEA technique is presented and the DEA model can be designed by deriving suitable factors for measuring university library efficiency through the review of related prior research. After this process, the university library(s) to be analyzed is selected, and the data from the library(s) are applied to the DEA model to measure each productivity.

Next, after evaluating the derived productivity and analyzing the difference in productivity between libraries, the cause and future tasks will be identified and provided as basic data for establishing strategies for improving performance.

This is consistent with the implications that this study intends to present. In addition, in order to present a more useful strategy for each library, a matrix that links the derived productivity with the library's performance is constructed to present the current position and strategy for each library.

2 Theoretical Background

2.1 Assessment of Service Efficiency

Measuring efficiency has attracted a lot of attention as a topic related to improving the productivity of an organization [14]. In the service industry, productivity improvement can be interpreted differently depending on what formula is used to define productivity, but basically it can be based on the ratio between output and input factors [15]. In other words, producing more output factors with the same input factor or producing the same output factor with fewer input factors can be said to improve productivity.

However, as suggested by Reichmann [1], in the case of university libraries to be analyzed as the focus in this study, there are a number of input factors and a production form of a number of output factors. Therefore, calculating efficiency and evaluating productivity as a ratio of a single pair of inputs can be limited. Therefore, applying a more suitable efficiency measurement method to the university library organization is the first step in the purpose of this study, and it will be able to contribute to the establishment of a strategy for improving productivity for the purpose of research.

2.2 Data Envelopment Analysis

Although multiple efficiency techniques can be applied to the efficiency evaluation of the service industry, Kim [8] stated that DEA is a linear plan developed to evaluate the effectiveness of non-profit and public decision making units (DMUs) whose causal relationship is not clear. Multiple efficiency techniques can be applied to the service industry's efficiency assessment. However, Kim [8] stated that DEA is a linear planning technique developed to evaluate the efficiency of non-profit and public decision making units (DMUs) whose causal relationship is not clear. It is a way to compare the output and input of DMUs, or to compare specific DMUs with other DMUs to show their relative (in) efficiency and size. Accordingly, DEA may be appropriate for university libraries, which are non-profit organizations, but preceding research studies are additionally reviewed to determine the suitability of DEA in this study.

According to the results of a study that reviewed a total of 34 previous studies related to the efficiency evaluation of university libraries, PLS (Partial Least Square)-SVM (Support Vector Machine) model, AHP (Analytic Hierarchy Process) model and various techniques were used, but among them, DEA model appeared as the most frequently used technique [16].

According to Tavares et al. [16], DEA can be applied to various situations and environments at the same time as the structure of multiple inputs and outputs is suitable unlike other techniques. They also evaluated it as a powerful tool for mathematically estimating the efficiency that could contribute to improving library performance. In addition, since there are many cases using DEA to measure the efficiency of domestic university libraries, DEA is considered a suitable technique for this study [17, 18].

As described above, university libraries have a production structure composed of a number of input and output elements. In this case, the DEA model cannot be constructed only with the ratio of output units to specific input units. It is possible to measure the efficiency of the central DMU by calculating all the detailed elements of the input unit (or the output unit) as a single result value through a weighted average, or to evaluate the efficiency through comparison with other DMUs [19].

2.3 Input and Output Variables

Determination of input and output variables is recognized as a very important factor in the use of the DEA technique. This is due to the research results which may appear differently depending on the selection of variables [20]. Therefore, the input and output variables used by reviewing previous studies that measured the efficiency of university libraries are summarized as shown in Table 1 below.

Table 1. Comparison of input and output variables in previous research

Reference	Input variables	Output variables
Han and Hong (2002)	Area of library space, Number of books Number of staff	Number of books used, Number of users
Reichmann (2004)	Number of books Number of full-time staff	Book materials added, Number of serial subscription, Regular opening hours per week, Total circulation,
Seo (2011)	Area of library space, Data purchasing expense, Number of books, Number of staff	Number of books used, Number of users, Number of book growth per year, Total number of serials
Jung (2011)	Area of library space, Data purchasing expense, Number of books and periodicals Number of seats, Number of staff	Number of borrowers, Number of library visitors, Number of loan books

(*continued*)

Table 1. (*continued*)

Reference	Input variables	Output variables
Park and Rhee (2016)	Area of library space, Data purchasing expense, Number of books, Number of enrolled students, Number of original text DB construction, Number of students for offline education students, Number of online education students, Number of reference service events provided, Number of staff	Corresponding DB search count, Number of homepage visitors, Number of library visitors, The number of loan books
Tavares et al. (2018)	Acquisition of books and periodicals, Area of library space, Book Collection, Total serials/Number of journals, Staff related variables	Book Circulations/Loan Transactions, Collections/Collection cost, Reader Visits/Number of users, Registered Readers, Total interlibrary lending

There are two points to note about the contents of Table 1. The first is that the variables used in the study by Park and Rhee [20] include the library's online service and electronic media. It is necessary to consider the influence of factors such as online, electronic media, and technology before selecting variables that are increasing in importance of ICT technology and non-face-to-face services. Second, Tavares et al.'s research [16] did not directly measure the efficiency of libraries, but rather reviewed and summarized related prior studies. In Table 1, the five input and output factors that were found to be the most used are listed.

Obviously, there are a number of studies examining the efficiency of universities with DEA, but it turns out that only Park and Rhee's research [20] focused on the digital elements of the DEA, such as database of university libraries, online education, and visits to the homepage. Besides, Noh [21] also conducted valuable study in respect to digital library by identifying new factors for evaluating efficiency of libraries in the digital era. The digital factors examined in this study are number of e-journals, e-books, computers, DB constructed, homepage visited and web DB used as well as sort of web DB. Due to the lack of previous studies on digital factors, this study has attempted to obtain those factors from valid sources of information, and the variables for DEA derived will be discussed in next chapter.

3 Research Method

In this study, the library to be analyzed were selected as follows. This study selected university libraries belonging to Group A among the groups classified by the Korea Education and Research Information Service (KERIS) based on the size of the universities for the DEA analysis. As the central analysis object, a library of University I, a four-year private university located in the Seoul metropolitan area, was selected. The reason for the selection of this library is that it focuses on providing online and non-face-to-face services such as delivery of copies of library materials and books for loans, and e-book use activation events, and it is relatively easy to secure data necessary for DEA analysis. On the other hand, for comparison, 25 private university libraries in the Seoul metropolitan area (Seoul, Incheon, and Gyeonggi) were selected. This is based on a study on the use of DEA that similarly established regional characteristics [22] and characteristics of operating subjects [23] between the central object and the comparison object.

3.1 Data Collection

For the selection of input and output factors, the following factors were considered mainly. According to the trend of digitalization and non-face-to-face service of library services in preparation for the era of the 'post corona,' input and output factors related to the libraries' online services was selected. In addition, some of the factors used to measure the efficiency of offline library services in previous studies [7, 21] were also applied. As a result, the input factors of university libraries to be used for DEA are the number of enrolled students, the number of staffs (NOS), the number of cases to establish digital materials (NDM), and the number of both Korean and foreign e-books. Here, the number of enrolled students is the sum of the number of undergraduate students (NUS) and number of graduate students (NGS), and the number of employees includes both permanent and temporary employees.

Digital content refers to electronic materials such as text and images built by university libraries. E-books are e-books stored on the library server or permanently stored. E-book refers to an e-book stored on the library server or permanently stored, and it is divided into domestic e-books (K) and foreign e-books (F). Since the difference between domestic and foreign book holdings for each DMU for the number of E-books to be described later is too different, it is used as a separate input factor. On the other hand, the number of Online Public Access Catalog (OPAC) searches and the number of digital data use (NDU) are selected as output variables to be used to measure the efficiency of university libraries. OPAC is a list of libraries that provide data search and communication functions online [24], and the number of searches is calculated by the number of times the search button is clicked on the library homepage. The number of electronic data use is the sum of the number of domestic e-book loans and the number of commercial DB use. The reason for using these factors in DEA is that KERIS' library statistics reflect the number of electronic data, including the number of e-books from overseas, as the number of commercial DBs. To this data, the number of e-books from Korea is added to create a complete number of electronic data usage. The input and output factors of the university library as above are selected centering on the library's online service. It is expected that it will be possible to derive results including offline service efficiency from a traditional point of view. The non-face-to-face service of the library, such as the delivery of new loan books after COVID-19, is not considered as a factor for using DEA because the actual number of cases was not large and is temporarily conducted within different periods.

Based on these research targets and the selection of input and output factors, we intend to construct a DEA model. STATA (Statistics/Data Analysis) 12 is used for the analysis of DEA. STATA is a statistical analysis software developed by Lee and Ji [25], and has been used for efficiency measurement and evaluation studies applying DEA [26, 27].

4 Data Analysis

This study has used the latest data provided by KERIS' Rinfo Academic Information Statistics System (Rinfo System) for DEA analysis [28]. The Rinfo system makes the provision of useful academic resources as its main business, and as part of that, it provides statistical data and statistical analysis reports of national university libraries every year. Statistical data used in this study was prepared as of 2019, of which only the number of enrolled students corresponds to the 2018 standard. The statistics on the extracted input and output factors are summarized in Table 2 and Table 3, respectively, and I University, the central object of the study, is DMU 1.

Table 2. Descriptive statistics of university libraries (N = 25)

Factors	DMU 1	AVG	StdDev	Max	Min
Number of students enrolled	20289	19724.84	6520.86	33487	10388
Area of library (m^2)	26218.1	25578.36	13197.65	68830	11575
Area of library per student (m^2)	1.3	1.28	0.42	2.3	0.6
Number of books	1659875	1600237	546656.70	923772	2985250
Number of books per student	81.8	82	14.34	115	52
Number of books borrowed	131719	144198	85478.55	333403	12879
Number of books borrowed per student	6.5	6.89	3.04	11.7	1.2
Number of visitors	1639932	2005267.95	1501691.14	6145352	255500
Number of visitors per student	80.8	96.20	51.49	222.5	24.6

Table 2 above shows the common descriptive statistics used as variables in previous studies [17, 18] on the library efficiency of all DMUs. With the difference between minimum and maximum values, and standard deviation values shown above, it appears that every university library has a distinctly different level of input and output variables used to assess the effectiveness of existing libraries. For DMU 1, compared to descriptive statistics, this library has slightly more students, area, and books, and less output from books and visitors.

Table 3. Descriptive statistics of each input and output variable

Factors	DMU 1	AVG	StdDev	Max	Min
NUS	17370	14400.24	4588.58	26063	6808
NGS	2919	3956	2576.43	11565	916
NOS	23	24.28	12.13	56	9
E-book (K)	54019	50123.72	41749.37	177990	8394
E-book (F)	33446	71119.44	77483.54	334273	0
NDM	65659	1810715	150902.25	154013.96	5646
OPAC	1465147	1273544.4	1203453.38	4716354	88375
NDU	2583841	7985854.36	10500975.11	51069461	285156

Table 3 above shows descriptive statistics for minimum and maximum values, mean and standard deviation for input and output variables. It is easy to see that all differences between minimum and maximum values, as well as the standard deviation value in each variable, are obvious. Compared to the average input, the central analysis, DMU 1, the number of graduates, the number of faculty members, the number of digital materials, and the number of overseas e-books and digital materials are relatively low while the number of undergraduates and e-books in Korea is higher than the average.

Statistics on the results show that students at DMU 1 are using the search engine of the school library relative to the average, but as opposed to using digital materials. The deviations shown in Table 3 made it difficult to determine the position of DMU 1 among 25 school libraries, so a DEA test was performed with 6 input variables and 2 output variables to investigate the actual efficiency of the selected library group.

5 DEA Result Analysis

Table 4 shows the results of DEA efficiency analysis for each university library.

Table 4. DEA result

DMU	Rank	Theta	Ref (DMU 1)
1	17	0.562	
2	1	1	0.0160393
3	10	1	
4	1	1	
5	12	0.997567	
6	19	0.344752	
7	1	1	

(*continued*)

Table 4. *(continued)*

DMU	Rank	Theta	Ref (DMU 1)
8	22	0.284403	
9	15	0.650265	
10	24	0.127081	
11	1	1	0.182854
12	23	0.13614	
13	1	1	
14	1	1	
15	20	0.3224401	
16	25	0.0645776	
17	16	0.629771	
18	1	1	
19	14	0.846668	
20	18	0.433955	
21	13	0.847159	
22	1	1	
23	21	0.297416	
24	11	1	
25	1	1	

According to Li and Ji [25], STATA conducting DEA provides multiple useful outcomes for testing respective efficiency of multiple DMUs. Theta in the table is equivalent to efficiency score and 1 indicates most efficient DMUs whereas 0 is the least. The term of Rank shows the position of each DMU for its efficiency. Then Ref in the table reports how a DMU, which is DMU 1 in Table 4, is referent to efficient DMUs. The larger Ref value of an efficient DMU displays the closer direction of future development for central DMU more efficiency. The DEA results performed in this study show that the efficiency of the metropolitan university library is uneven.

Twelve libraries can be identified as top group (where Theta > 0.9). The nine libraries are highly efficient where three additional libraries (DMU 3, 5 and 24) appear very close to rank 1. Then five libraries can be placed in the middle group (where Theta > 0.5) and DMU 1 is one of this group. Lastly, the remaining eight DMUs can be arranged in a subgroup (where Theta < 0.5) and the most noticeable result in this group is DMU 16 with a very low efficiency score.

All libraries with Theta lower than 1 seem to have needs to improve their efficiency in operation. While Ref indicates the need for efficient DMU benchmarking, DMUs 2, 11 and 14 can be top three best practices these DMUs are most relevant to other DMUs. To be specific, DMUs 2, 11 and 14 can be best practices for DMUs 7, 16 and 5. These results also show that DMU 11 appears to be a very efficient university library in the metropolitan area, while DMU 11 can be benchmarked on sixteen other DMUs. In the case of DMU 1, the library is able to benchmark DMU 2 and DMU 11 in the aspect of inputs and outputs used by DEA, and has a positive Ref value for efficient DMUs. Thus DMU 11 with highest Ref value could be best practice for DMU 1. However, it is difficult to pinpoint the exact area, but only the Theta and Ref and Slacks, which are additional results derived from DEA by STATA, report the relevant information.

Table 5. Slacks for input and output variances

DMU	Input slacks						Output slacks	
	NOU	NOG	NOS	e-book (K)	e-book (F)	NDM	OPAC	NDU
1	6982.29	1168.47	8.9321			28054.4	0	6789672
2		0	0	0		0	0	0
3	2014.39	344.177	5.81167	11766.7			0	3406050
4		0	0				0	0
5		603.812	0.312432	25536.9		15550.6	0	37927.4
6	4668.05	1931.32	5.38169			3456.01	0	2906011
7		0	0		0		0	0
8	380.61	276.313	3.39042	24187.4			0	2664407
9	2756.9	700.921	3.98394			27608.9	0	1400071
10		216.366	0.155433	312.352		5612.42	0	125247
11		0	0	0		0	0	0
12	525.825	69.715	0.634922	5830.43	1650.63		0	508287
13		0	0				0	0
14		0	0		0	0	0	0
15	2982.89	613.912	2.295529	546.676	2251.23		43420.1	0.007753
16	470.339	34.2058	0.696019	1963.55	2013.83		0	246138
17			0.437806		9233.44		0	3244139
18			0		0		0	0
19	1653.62	2706.62	8.14191			4203.57	0	3917387
20		1922.73	5.10995				184353	0.169263
21	1117.15	806.327					33743.8	
22			0		0		0	
23	3344.47	343					0	3048591
24	2613.55	4021.26					0	7603096
25			0		0		0	

The slacks, accompanying with Ref value, is a concept providing the information regarding the source of inefficiency [25]. A DMU with Theta score of 1 and no slack values (slacks of 0 indicates not fully utilized) derived can be identified as fully efficient [29]. In this regard, nine DMUs (DMU 2, 4, 7, 11, 13, 14, 18, 22 and 25) are fully efficient through our DEA whereas DMU 3 and 4, despite their Theta, are excluded from this group. Agarwal et al. (2011) also suggested that input slacks is equivalent to the amount of input excess could be reduced with the same output and output slacks, in similar, indicates the amount of output shortfall.

Based on these ideas, a series of beneficial results for inefficiencies are derived as the results in Table 5. First, number of graduate students seems to be the most noticeable problem among DMUs regarding input variable with twenty-one DMUs with high Slacks values. Second, the number of staffs is derived as the most common problem, and twenty-two DMUs have Slacks values. However, it seems not to be a significant problem with all slacks on this variable are comparatively low. Third, for the output variables, number of digital materials used can be major issue for efficiency due to the twenty-two DMUs with tremendous Slack values. Lastly, DMU 1 is as efficient as the number of e-books in the country and abroad. The next prominent inefficient variables are number of undergraduate students, the established digital materials, and digital materials used.

The DEA results show a number of inefficient factors for all DMUs with Slack values, but are limited in proposing real solutions, especially for input variables.

For instance, it would not be practical to reduce number of undergraduate students and digital materials merely for input excesses. Keeping with this idea, DEA test-based solutions can be output oriented. For DMU 1, for example, the main suggestion for improve efficiency are: While number of students is inevitably fixed, DMU 1 can reduce the number of unpopular digital materials and replace them with specific materials tailored to the users' needs. This solution should be accompanied with a survey about users' needs for the library's digital material. More customized digital materials will increase the number of being used, reduce production and increase efficiency. In addition, benchmarking also can be an effective solution for inefficient DMUs. For each inefficient DMU, it is able to target any DMU for best practice based on Ref value and investigate how to operate aspects with slacks in efficient way.

6 Conclusions

This study focused on the efficiency assessment by conducting DEA. For DEA, firstly, the study selected 25 libraries from four-year private universities located in the Seoul metropolitan area. The main issue on the DEA was to examine efficiency of the libraries in respect to online and digital service factors. Thus the relevant input and output variables were selected. For input variables, numbers of students enrolled and library staffs were selected as traditional ones whereas number of e-books and cases to establish digital materials reflect online and digital aspects. Two output variables, on the other hand, of the number of OPAC searches and the number of digital data used were selected to measure the output of digitalized and web-based library services.

The result of DEA showed that 9 libraries are measured as fully efficient DMUs while other 16 libraries are not. In addition, the efficiency score of each library is widely varied while 8 of libraries are under Theta value of 0.5 as well as the least score was 0.0645776. For further analysis, the reference and slack scores are used. As a result, several useful implications were derived. First, it shows inefficient in input variables in number of graduate students and that of digital materials established for majority inefficient libraries. Certain strategy for encouraging more students to use digital materials could be suggested for this problem or simply reduction of numbers for e-books could also be considered merely on the view of efficiency. Second is mainly about the central analysis target, DMU 1. Libraries need to resolve the inefficiency of undergraduate students and digital materials that are established and used, for instance by reducing unpopular materials while providing new personalized materials. In addition, DMU 1 offers more alternatives to benchmarking efficient DMUs which is highly relevant shown as DEA result of reference score.

This study is one of the first trial to assess the effectiveness of libraries focusing on digital and online services through DEA testing and is able to derive valuable implications. However, there is a limitation in that is not possible to obtain recent data on non-face-to-face services developed in response to corona spread and more practical guidelines for library improvement in the Post-Corona era are being presented. In addition, it is unable to provide solutions details because the data represents only the amount of variables and does not specifically show how and the quality of delivery to clients as a form of library service. Based on these ideas, future study will focus on new services formed after COVID-19 and the best practices of university libraries derived from this study to provide guidance for improving efficiency and adaptation to Post Corona.

References

1. Reichmann, G.: Measuring university library efficiency using data envelopment analysis. Libri **54**, 136–146 (2004)
2. Wang, M.Y.: Introducing CRM into an academic library. Libr. Manag. **28**(6), 281–329 (2006)
3. Sahu, A.K.: Measuring service quality in an academic library: an Indian case study. Libr. Rev. **56**(3), 234–243 (2006)
4. Islaml, M.A., Agarwal, N.K.K., Ikeda, M.: Knowledge management for service innovation in academic libraries: a qualitative study. Libr. Manag. **36**(1/2), 40–57 (2015)
5. Islaml, M.A., Agarwal, N.K.K., Ikeda, M.: Conceptualizing value co-creation for service innovation in academic libraries. Bus. Inf. Rev. **32**(1), 45–52 (2015)
6. Kim, S.A.: Evaluation efficiency of university libraries through data envelopment analysis. J. Korean Libr. Inf. Sci. Soc. **38**(2), 137–160 (2004)
7. Seo, S.D.: The analysis of efficiencies and productivity changes of university libraries in Korea. J. Ind. Econ. Bus. **24**(1), 47–71 (2011)
8. Kim, Y.S.: A study of service operations efficiency evaluation in Korean banks. J. Manag. Econ. (Inha Univ.) **16**(1), 145–164 (2002)
9. IFLA: COVID-19 and the Global Library Field (2020). https://www.ifla.org/covid-19-and-libraries. Accessed 09 Oct 2020
10. Committee of Korea for Information: Policy of Library, the Implementation Plan of the 3rd Comprehensive Plan for Library Development (2020). http://www.clip.go.kr/main/PageLink.do. Accessed 10 Oct 2020
11. Shin, H.S.: Lending Books to New Enrolled Students. Inha University (2020). http://www.dhnews.co.kr/news/articleView.html?idxno=119754. Accessed 11 Nov 2020

12. Lee, D.S.: Books are sent on delivery. Yeungnam University library (2020). http://news.unn.net/news/articleView.html?idxno=228275. Accessed 11 Nov 2020
13. Baek, D.S.: Open the First Smart Space for University Library. Yonsei University. http://www.dhnews.co.kr/news/articleView.html?idxno=117665. Accessed 11 Nov 2020
14. Cook, W.D., Seiford, L.: Data envelopment analysis (DEA) - thirty years on. Eur. J. Oper. Res. **192**(1), 1–17 (2009)
15. Rutkauskas, J., Paulavlelene, E.: Concept of productivity in service sector. Eng. Econ. **3**(43), 29–34 (2005)
16. Tavares, R.S., Drumond, G.M., Angulo Meza, L., Mexas, M.P.: Efficiency assessment in university libraries. Transinformação **30**(1), 65–79 (2018)
17. Han, D.O., Hong, B.Y.: Efficiency of university libraries in Seoul by DEA. J. Korean Libr. Inf. Sci. Soc. **36**(3), 275–286 (2002)
18. Jung, Y.M.: Evaluation of university library efficiency using data envelopment analysis. J. Korean BIBLIA Soc. Libr. Inf. Sci. **22**(4), 301–315 (2011)
19. Kim, S.B., Soh, S.B.: Measuring and comparing the performance of local museums using DEA technique. Korean Soc. Sci. Art **27**, 57–67 (2016)
20. Park, H.Y., Rhee, H.Y.: A study on the operational efficiency analysis of university libraries. J. Korean Soc. Inf. Manag. **33**(1), 139–160 (2016)
21. Noh, Y.H.: A study on identification of library resources for evaluating efficiency of academic libraries. J. Korean Biblia Soc. Libr. Inf. **21**(1), 71–88 (2010)
22. Joo, H.S., Lee, K.M., Jung, C.M.: Efficiency analysis for public library with data envelopment analysis in Seoul. Seoul Stud. **15**(2), 131–149 (2014)
23. Kim, S.A.: Comparing the efficiency of public libraries. J. Korean Soc. Libr. Inf. Sci. **41**(2), 237–256 (2007)
24. Rhee, H.Y.: A survey of user perceptions of OPAC 2.0 service in academic library. J. Korean Libr. Inf. Sci. Soc. **43**(2), 179–201 (2012)
25. Lee, C.J., Ji, Y.B.: Data envelopment analysis in Stata. In: Stata Conference (2009)
26. An, J.H., Choi, B.C., Kim, S.B.: A study on relative management efficiency of domestic coffee franchise companies using data envelopment analysis. J. Foodserv. Manag. Soc. Korea **22**(1), 365–379 (2009)
27. Lee, J.P., Lee, S.B., Choi, B.C.: Measuring the relative efficiency of art museum in utilizing social network services: DEA approach. Korean Soc. Sci. Art **38**(1), 169–179 (2020)
28. Rinfo Academic Information Statistics System. http://www.rinfo.kr/. Accessed 11 Nov 2020
29. Agarwai, S., Yadav, S.P., Singh, S.P.: A new slack DEA model to estimate the impact of slacks on the efficiencies. Int. J. Oper. Res. **12**(3), 241–256 (2011)

A Study on Airport Service Improvement Using Service Design Process

Seo Young Kim[1], Tae Hee Kim[2], Youn Sung Kim[1(✉)], and Min Seo Park[1]

[1] College of Business Administration, Inha University, Incheon, Korea
sysb012@hanmail.net, motbeol@inha.ac.kr, 22201580@inha.edu
[2] Airport Passenger Service Incheon, Korean Air, Incheon, Korea
taeheekim@koreanair.com

Abstract. The aviation industry is expanding non-face-to-face services using self-service technology. In addition, major airports around the world are rushing to introduce 'smart airport' in order to maximize customer satisfaction as well as responding appropriately to air demand. Therefore, it is significant to focus on this innovation in aviation industry. The purpose of this study is to analyze changes in airport service environment and process, to derive improvements on pain point found through in-depth interviews and persona production of employees and customers, and to establish an improved airport service process and service environment by reflecting the needs of customers. There are academic and practical implications in that it presented a convergent idea to provide balanced service to non-face-to-face service preferred.

Keywords: Airport service · Non-face-to-face services · Service design process · Persona · Customer journey map · Service blueprint

1 Introduction

The aviation industry is expanding non-face-to-face services using self-service technology due to the advantages of reducing the burden of customer response deviation and employee wage payment, which are disadvantages of face-to-face services, and more efficient and economical in terms of customer waiting time and opportunity cost for convenience [1].

In addition, major airports around the world are rushing to introduce 'smart airport' in order to maximize customer satisfaction with customized services as well as the purpose of responding appropriately to air demand. In other words, companies have been able to reduce fixed costs and provide more efficient services through the simplification and computerization of business procedures by using technology-based self-service, which is a non-face-to-face service, and customers have been able to use aviation services more effectively and conveniently through various routes, away from time and space constraints [2].

The aviation industry, which has been labor-intensive and face-to-face service, has expanded non-face-to-face service technology to many areas such as reservation, ticketing, and airport service after the introduction of e-ticket in 1994. With the convergence

R. Lee (Ed.): ICIS 2021, SCI 1003, pp. 75–85, 2022.
https://doi.org/10.1007/978-3-030-90528-6_7

of ICT (Information and Communications Technology) technology, customers use web or mobile devices on their way to their home or airport instead of going to the counter where they can meet employees and, it is possible to check in with the kiosk installed at the airport. If a reservation change occurs, the customer will be able to adjust the schedule in response to the system. The automatic registration tag system for the baggage consignment will be available, and when the baggage is lost, the customer will be able to access the baggage by himself and track the location of the cargo.

In fact, the International Air Transport Association (IATA) recommended each airline to introduce a Radio-Frequency Identification baggage tracking system by 2020, and the 75th IATA Annual Meeting held in June 2019 decided to support the global introduction of the Air Log tracking system (applying RFID-based technology). The automatic travel document review will be carried out by using the passport automatic recognizer, and when boarding, the decision to open and close the boarding gate will be made by communicating directly with the customer by the system while having the position of the boarding bridge through the mobile device and the information of use. These processes are expected to be available at smart airports in accordance with the characteristics of the airport users without being cut off step by step [3].

Thanks to this atmosphere, Korean Air, a large domestic airline, announced on September 1, 2019 that it will convert and operate the general seat counter of domestic airports into a baggage counter [4]. In other words, all customers on the general seat must issue their boarding pass directly through non-face-to-face services such as the Internet, mobile, or kiosks at the airport.

The purpose of this study is to analyze changes in airport service environment and process, to derive improvements on pain point found through in-depth interviews and persona production of employees and customers, and to establish an improved airport service process and service environment by reflecting the needs of customers.

2 Theoretical Background

2.1 Trends in Non-face-to-face Services in the Aviation Industry

Non-face-to-face channels are channels that allow customers and employees to select and purchase products through various information and communication media or internet networks without face-to-face, which is linked to services in various industries as new technologies develop due to rapid information and communication media or technology development [5]. As this new technology is combined, services can be divided into face-to-face services and non- face-to-face services according to the objects interacting at the service contact point.

The aviation industry, which has been dominated by face-to-face services, is investing in technology to reflect the trends of the times and to improve the efficiency of company operations. Self-Service Technology (SST), a non-face-to-face service, is widely implemented in the aviation industry, and the biggest change in the aviation industry is the formation of customer-led self-service technology [6, 7].

The representative types of airline self-service are web-based, mobile app-based, kiosk-based, and self-backdrop-based four types. The technology to check the identity through facial recognition without showing travel documents and board through the unmanned smart gate is in the introduction stage at Incheon International Airport and is in the process of commercialization soon. In order to survive the fierce competition with airports in each region pursuing smart airports, airlines will continue to expand the state-of-the-art self-service area where ICT technology is fused.

2.2 The Concept and Process of Service Design

Service Design began in 1984 when Shostack proposed the concept that service can be designed through service blueprint that can categorize the service of intangible characteristics as design, and the UK's "Engine" and the US's "IDEO" are spreading as major projects are being carried out [8–10].

The UK Live:Work company defines service design as 'designing various contact points that people encounter over time so that customers can have diverse experiences that are not uniform', and the US Engine Group defines 'developing each element to make the service easy, satisfying, and efficient by covering various fields of design such as environment, communication, and products' (this is the case) [11–13].

Service design is going through the process of discovering and solving problems through design thinking, which can be said to start from empathizing with users at future point of view rather than present. It starts with an understanding of the details such as what you want, what you need, what you prefer [14].

The service design process is composed of 6 steps or more from 3 to more than 6 steps depending on domestic and foreign design companies, and the terms by step are slightly different [15, 19] (Table 1 and Fig. 1).

Table 1. Main service design process

Company	Process
Live work	Insight → Ideas → Prototype → Delivery
Design Council	Discover → Define → Develop → Deliver
IDEO	Observation → Brainstorming → Rapid prototyping → Refining → Implementation
Engine	Discovering → Informing → Exciting → Optimizing → Specifying
Moritz	Understanding → Thinking → Generating → Filtering → Explaining → Realizing
Team interface	Discovering → Concepting → Prototyping → Specifying → Testing

In this study, 'Double Diamond Design Process Model', which is the most representative methodology among various service design processes, was used. This process is characterized by convergent thinking and divergent thinking, and the detailed process consists of Discover – Define – Development – Deliver [16, 17].

Fig. 1. Modified general seat counter

3 Design Process of Airport Service

3.1 Change of Airport Service Environment

The customer classification, which was divided into individual travelers and group travelers, is divided into passengers who have completed pre-processing and passengers who have not checked in advance due to the suspension of group counter operation and the reorganization of counter operation. From September 1, 2019, Korean Air's Incheon International Airport General Seat Counter has changed all of the following to baggage counters. For the passengers who cannot use kiosks because of the visa checking or another mistakes, Help Desk is additionally arranged and operated.

Customers who have completed pre-check-in through mobile or web check-in will send their luggage from the baggage counter after arriving at the airport and move to the departure gate. If you are a customer at the airport without checking in, you should use the kiosk to issue your boarding pass first. Although it operates a help desk, it operates only two, so it is inevitable to wait for a long time in the concentration time zone. In the concentration time zone, the mobile counter is opened around the kiosk to support the issuance of boarding tickets for passengers with exchange cards. The baggage only counter can be accessed by presenting a boarding pass or exchange card, and the counter does not provide services other than receiving the baggage. Group travelers should also check in as well as individual travelers, or issue boarding tickets individually through kiosks at the airport. Only up to six people can check in at the same time using kiosks. In this process, it was confirmed that the items changed to dissatisfaction factors among the existing satisfaction factors were 'One Stop Service' and 'Waiting Time'.

3.2 Persona Definition

Based on in-depth interviews with customers, field observations, and in-depth interviews with employees, a total of two personas were derived. The persona derived is 'Lee, Sul-baek', a 20-year-old female passenger who prefers non-face-to-face services, has experience studying abroad, and enjoys SNS, and 'Na, Dae-Sin', a 70-year-old male passenger who has no difficulties in using IT but prefers face-to-face services and was retired as a large company executive (Table 2 and 3).

Table 2. Representative Persona 'Lee Sul-Baek'

	Name: Lee Sul-Baek Age: 26-year-old Average monthly income: 4.5 million won Gender: Female Residential: Seocho-gu, Seoul Married: Unmarried Job: Elementary school teachers Annual average number of trips: More than once Keywords: Restaurant, Mead, SNS, Physical Fitness, Experience, Challenge
Personal information	She studied at American University for a year as an exchange student. She had been backpacking alone in Europe when she was in college. She lives with her parents, saving 1.5 million won when she gets paid, and she saves some money and has a lot of money in her travel account. She often travels to Japan near her to buy duty-free cosmetics, and she has no boyfriend, and likes to leave a day's record on Facebook and Instagram. She downloads the mobile app, performs mobile procedures at home in advance, and goes to the airport and mainly uses self-back drop counters.
Needs	She hates the many people and the complexities. She wants to check in on mobile devices and then handle the luggage quickly and easily without face-to-face with employees at the airport. She wants emergency exit seats to be able to be checked on the Internet or on mobile, even if she pays extra. She wants to have information gathered in one place because it is not easy to prepare for a trip in a busy life .She wants to be taken immediately on the spot if there is a problem with using the kiosk, and it is uncomfortable to be guided by a complicated process.

3.3 Customer Journey Map

1. The story of Lee Sul-Baek

This trip was to travel to Europe with close friends and decided to check in on mobile at home in advance. I thought I wanted to sit at the exit and relax on my way, but I couldn't book or check the exit seat on my mobile, so I chose the seat and finished the procedure. When I arrived at the airport, I went to the B counter, which is a counter for excellent members. I am going to enter, but only one person can enter. I think you could just let me in, but I'm not happy to be rejected as soon as I arrive at the airport. It feels like it's the end and the end.

Table 3. Representative Persona 'Na Dae-Sin'

	Name: Na Dae-Sin Age: 70-year-old Family: 2 boys and 1 girl Gender: Male Average monthly income: 5 million won Marriage: Married (40 years of marriage) Residence: Mapo-gu, Seoul Jobs: Business Managing Director of Large Enterprises (now retired) Annual average number of trips: More than twice Keywords: Health, Family, Relationships, Travel, Comfort and Familiarity
Personal information	It was strange at first to leave the company that had been in for more than 40 years, and to get up every morning and to have nowhere to go. The youngest daughter lives near the house and often comes home with her daughter. He is happy to have a lot of laughing and doing things with his granddaughter, and He regularly travels abroad with his wife at circle. He often reads books in the nearby bookstores or libraries. He pays his dues and travels regularly in couples once a year, and sometimes to Southeast Asia for a golf, and he travels in domestic places with his family.
Needs	It is annoying and uncomfortable for him to know and prepare for his travel. He expects the appropriate service of travel agency or airline staff. He remembers that over weighted-burdens and reorganizing burden made him exhausted. He wants quiet place to prepare it. He thinks it's a luxury to ride first class, but he wants once if he could.

I try to get into the G counter and show me my boarding pass. I panicked. I checked the mobile, but I think it is possible to issue a boarding pass to Kiosk. I have to move to the emergency exit seat anyway, so I go to the kiosk and wait and try to use the kiosk. I cannot scan my passport and it keeps getting error. I tried it several times, but it didn't go on. The first button was important. I finally got to line up the waiting line after I talked about it. By the way, there are many counters and there are not many people waiting for us, but there are a lot of people waiting for the help desk where we are, and only two counters. I waited more than 20 min and I went to the staff, but the emergency exit seats are not gone.

I had been preparing for Europe for a few months because I had traveled to Europe with my close friends, but I arrived at the airport and waited for me to stand in line from the beginning. I waited for the kiosk to hold the kiosk machine and wrestled with it. I am already tired before the start of the trip.

2. The story of Na, Dae-Sin

It's a couple's travel. I arrived at the airport with a golf club. I lined in the nearest B counter. But it's for the business class. The staff said to me that I should go to G Counter. I wanted to upgrade with my mileage and get a business class, but I didn't. I moved to G Counter thinking I wasn't satisfied. When I tried to line in the counter with the staff, the staff stopped me and he said that all the counters were operated for a baggage-loading, so he asked me to use the front device (kiosk), to get the boarding pass. I also wanted to check if I could upgrade with my mileage. But when I went to the kiosk as I was told, I knew that I had to try by myself, because all the staff was very busy to guide other person.

I scanned my passport and it said I didn't have any reservation information. I searched it in the itinerary that the travel agency gave me. Where is the reservation number? If I go to the flight attendant, I submit my passport, he could ask me what seat I wanted and aisle seat could be O.K? There I met my fellow men. They all want to use the kiosk together, but it was very difficult. Nobody knew what do. We asked the staff over there and he said that we could be allowed to check only six people simultaneously. As a result, two boarding tickets were not printed out, but the staff didn't know the reason why.

We were waiting in the help desk line, but there were more people than the baggage counter line. When I asked if we could go to the side line, the employee answered that passengers without a boarding pass couldn't enter. I thought that he was not flexible. I really want the staff to check in and load up as they did before. It was said that it was more convenient and faster to use kiosk, but it took more time because the group was not able to do it together.

After all, the person who wanted to ask if he could upgrade with a mileage, and the person who could not take the boarding ticket from kiosk, should be in line again. They should wait for a double time, so I thought that it was more uncomfortable system.

3.4 Service Concept and Improvement

The first, the Payne Point, which can be known through the customer journey map, was mobile/web check-in, and it was impossible to make reservations for emergency or some seats in the pre-check. In addition, there was a hassle for customers to find detailed information by menu and lack of information about airport use.

Second, since the counter was subdivided after arriving at the airport, it was inconvenient to move to the available counter by customer characteristics rather than the nearest counter, and the advance notice about the counter operation was not on the Internet. Third, all of the customers who did not check in in advance should check in using the kiosk, but there are many cases where they have to visit the counter with the staff because the kiosk function improvement is not done together due to the change of service environment.

Fourth, customers who do not want to use the kiosk, customers who need to check their visa, and customers who need to receive service from the kiosk because they cannot output their boarding pass, have to line up with a help desk.

Through the analysis of service environment and process, interviews with customers and employees, persona, and Payne points of customer travel guidance, three service concepts (Seamless Self Service, One Stop Service, and Waiting Time Management) were derived.

The specifics of the service improvement plan are as follows: First, it is the provision of a seamless self-service. When using aviation services, it is necessary to provide a service environment that can use non-face-to-face services from purchasing airline tickets to boarding airplanes. For this, sufficient information should be provided through mobile apps or internet homepages, which are the only information providing channels for customers, and if customers input customer characteristics rather than looking for them by themselves, they should improve the direction of providing the information necessary for customers directly by using big data.

Second, one stop service is provided. In the past, when I came to the airport with only tickets, passports, and luggage, the staff assigned the seats, checked the travel documents, and sent the luggage. However, with the full implementation of self-check-in services, there is room for customer complaints. If customers do not agree with it as more things are needed to do. Therefore, even if the self-service device is used, a one-stop service environment should be provided to receive help right there if a problem occurs.

Third, it is the management of waiting time. Relative waiting time management is more important here, and when there are customers who only load luggage at the baggage counter and customers waiting to use the help desk, the customers in the help desk waiting line often feel that the relative waiting time is long. Customers who are unable to use kiosks or need help from employees due to problems already have experienced more than one service failure, so they feel a long waiting time and actually have only two help desks. It is necessary to manage the waiting time so that it does not lead to customer complaints.

Some of the detailed improvements of 'Seamless Self Service' are as follows. It is to diversify the airport non-face-to-face service channel. It is to introduce 'Self-help desk' operation, 'Overload self-counter' operation, and 'Digital bag tag' service operation.

First, Self Help Desk has a device that recognizes the barcode near the kiosk so that customers who need guidance from the changed seats or employees can check their own boarding, seating, and on-time operation. If necessary, it makes it possible to send a mobile boarding pass to the customer's mobile phone (Fig. 2).

The 'Excess baggage self-counter' adds function to the existing electronic balance. Currently, electronic balances are provided near the airport counter so that customers can weigh themselves and arrange their luggage in advance before entering the counter. The Excess Cargo Self Calculator is to combine the self-calculator with the scale so that customers can arrange their bags themselves and pay for excess baggage. For customers, the weight is exceeded at the counter, which can reduce the hassle of re-arrangement. For employees, only the receipts already calculated and the baggage table are issued, so the time for the luggage check is shortened (Fig. 3).

The Digital Bag Tag is designed by applying the case of British Airways, which sells digital baggage tickets that can be reused more than 3,000 times for a fee. If you attach a personal tag to your baggage and work with the mobile app, a new baggage number will be created in the digital tag [18].

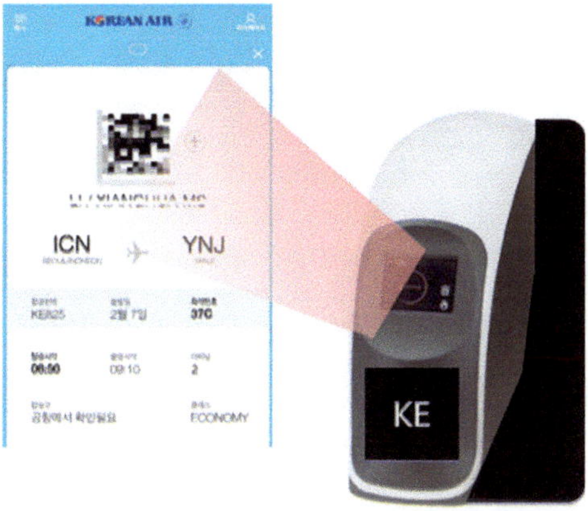

Fig. 2. Virtual self help desk

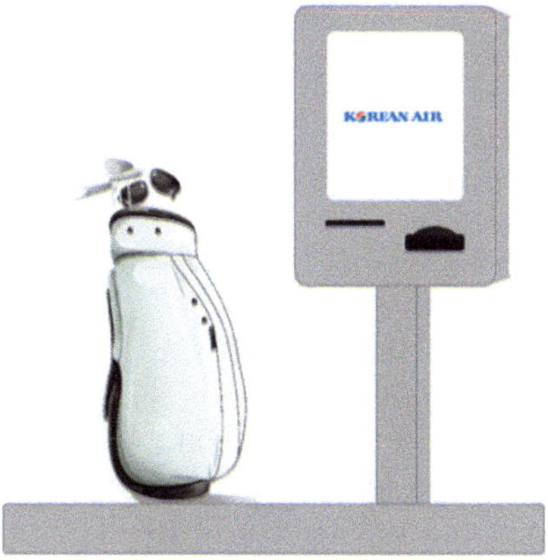

Fig. 3. Virtual excess baggage self-counter

You just have to come to the airport and go to the Tag Bag Drop unmanned counter and put your bag up. It is possible to track the baggage location even when the baggage is lost or delayed. The most worrying thing about self-service is security. Because I have my own choice, it will be effective in preventing accidents such as putting wrong choice or attaching other people's choice. If digital tag service is introduced, I expect that the era of 'Seamless Self Service' will be opened (Fig. 4).

If the above services are newly applied, there will be no contact between customers and field staff in the process of using airport services. Customers who prefer non-face-to-face service, can use mobile apps booking and checking in without any restrictions on seat selection, and then can move to the departure place without contact even when using the airport on the same day through digital bag tag service.

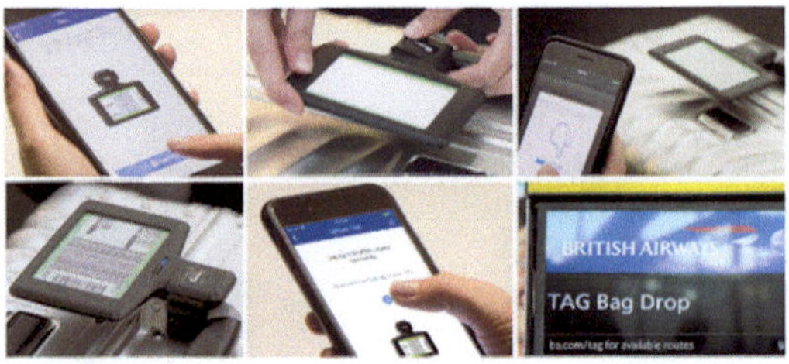

Fig. 4. British Airways' 'Digital bag Tag'

4 Conclusion

This study was conducted on Korean Air, which recently announced a full self-check-in service by changing the operation method of the general seat counter at domestic airports. In order to improve the boarding service environment and process of airport services, which are mainly non-face-to-face services, and to present future directions, service design was used to study.

Seamless Self Service can help passengers who prefer non-face-to-face services to complete the luggage process quickly without the help of employees at the airport through non-face-to-face services through diversification of airport non-face-to-face service channels such as self-overload checkout, digital bag tag service, and self-help desk installation and operation can make a service environment capable of making a service environment.

There are academic and practical implications in that it presented a convergent idea to provide balanced service to non-face-to-face service preferred customers and face-to-face service preferred customers, but it should be applied to the business through the demonstration process of the service prototype in the future and the results derived should be constantly revised and supplemented. In addition, in order to balance the needs of companies and customers, it is necessary to look at the overall aviation service process.

References

1. Kim, A.G.: A Study on the Influence of Airline Self Service Quality on Relationship Quality, Reuse Intention and Intention of Words-of-mouth. The Graduate School of Aviation & Management, Korea Aerospace University, Master's thesis (2019)
2. Lee, S.J.: Effects of Technology-Based Self Service on Experience Value and Satisfaction, Behavior Intention. The Graduate School of Service Management, Kyong Gi University, Master's thesis (2018)
3. Hyun, W.S.: Information and communication technologies for smart airport. Korean J. Aerosp. Environ. Med. **28**(2), 30–37 (2018)
4. Korean Air. www.koreanair.co.kr. Accessed 10 Nov 2020
5. Lee, S.Y.: A Study on the Service Quality of Online Channel for Insurance Purchase. Graduate School of Changwon National University, Master's thesis (2018)
6. Lee, S.M.: A Study on the Technology Acceptance of Airport Self-Service Technology Users. Graduate School of Korea Aerospace University, Master's thesis (2018)

7. Kim, E.H.: The Effect of Customer Readiness on the Use Intention of Airline Self Bag Drop System. The Graduate School of Aviation & Management, Korea Aerospace University, Master's thesis (2017)
8. Choi, J.M., Chun, Y.D.: A case study on healthcare service design - focused on service innovation current situation analysis in hospital. J. Korean Soc. Commun. Des. **17**(2), 197–207 (2014)
9. Choi, S.I., Kim, Y.S., Yoo, H.J., Jang, J.B., Hwang, J.H.: Service Management 4.0, Seoul, MoonWooSa (2016)
10. Shostack, G.L.: Designing services that deliver. Harv. Bus. Rev. **62**(1), 133–139 (1984)
11. Yi, S., Kim, B.Y., Seo, J.A.: Exploring the service design in the airline industry: a case of virgin Atlantic airways. J. Prod. Res. **28**(4), 79–95 (2010)
12. Livework (2018). https://www.liveworkstudio.com/monthly-magazines/4-secrets-to-successful-disruption-through-service-design/. Accessed 9 Nov 2020
13. Livework. https://www.liveworkstudio.com/. Accessed 9 Nov 2020
14. Ko, E.H., Kim, E.: Traditional market service design using design thinking process. Proc. Korea Soc. Des. Stud. Conf. **2018**(6), 130–135 (2018)
15. Suh, C.J., Lee, H.J.: A case study on applying service design methodology to improve the premium food market experience. J. Korea Serv. Manag. Soc. **19**(4), 201–220 (2018)
16. Kim, S.D.: Analysis of Medical Service Design Case in Terms of Patient Experience. Yonsei Graduate School of Human Environment Sciences, Master's thesis (2019)
17. UK Design Council. www.designcouncil.org.uk. Accessed 11 Nov 2020
18. British Airways. www.britishairways.com. Accessed 11 Nov 2020
19. Kim, C.S., Nah, K.: A study on the utilization of open innovation platform for co-creation as service design methodology: focused onthe case study of OpenIDEO and Quirky. J. Digit. Des. **13**(1), 333–342 (2013)

Effect of Empirical Value of Untact Marketing on Consumer Satisfaction and Repurchase Intention: Centered on Service Application

Jin-Hee Lee(✉)

Department of Business Administration, Soongsil Cyber University, Seoul, Korea
clara@mail.kcu.ac

Abstract. Untact marketing is increasing due to the worldwide COVID-19 pandemic. Untact is a new term that originated in South Korea combining "contact" and "un" which refers to non-face-to-face contact. Typical untact services include a kiosk, which refers to an unmanned cash register, AI chatbot service, and delivery mobile applications. Untact marketing is an optimized way for smartphones. Untact marketing is generating profits by communicating with consumers through non-face-to-face methods and by e-enabled marketing targets. As the changes in the marketing environment, an investigation was conducted on how the empirical value of untact marketing affects consumer satisfaction and consumer's willingness to repurchase, centered on service mobile application. This investigation intends to contribute to the establishment of untact marketing strategy. After using service mobile application that is frequently used in our daily lives, such as delivery apps, shopping apps, and finance apps, we studied consumer satisfaction with their empirical value and their intention to repurchase them and tried to present the direction for the strategic development of untact marketing by revealing the influential relationship. Therefore, empirical values were contemplated, and according to this, concepts of consumer satisfaction, willingness to repurchase were summarized. The research model was constructed based on the theoretical background, and the survey was conducted for empirical analysis. The collected data was obtained using the SPSS 22 program. In this research model, empirical values perceived by consumers using mobile service applications were defined as five factors: hedonic value, practical value, economic value, service value, and reliability value. In this study, consumers seem to be positive about untact marketing, and as the consumer market for the untact marketing expands, it is believed that corporate strategic efforts are needed so that consumers can feel interested, as well as realizing economical, and reasonable value. Consumers seem to be positive about untact marketing, and as the consumer market for the untact marketing expands, it is believed that corporate strategic efforts are needed so that consumers can feel interested, as well as realizing economical, and reasonable value.

Keywords: Untact marketing · Service application · Consumer satisfaction · Repurchase intention

R. Lee (Ed.): ICIS 2021, SCI 1003, pp. 86–95, 2022.
https://doi.org/10.1007/978-3-030-90528-6_8

1 Introduction

Untact marketing is increasing due to the worldwide COVID-19 pandemic. Untact is a new term that originated in South Korea combining "contact" and "un" which refers to non-face-to-face contact. The emergence of COVID-19 has changed our daily lives, and untact service is already existing in our lives. Typical untact services include a kiosk, which refers to an unmanned cash register, AI chatbot service, and delivery mobile applications. Untact marketing is an optimized way for smartphones. Untact marketing is generating profits by communicating with consumers through non-face-to-face methods and by e-enabled marketing targets. As the non-face-to-face consumption culture is widely established around the world, it is time to change marketing and methods of distribution [6]. With the surging number of single-person households, the population and generation structure are changing, and more and more people are trying to use the untact service due to changes in consumer attitudes. In particular, the lower the age group, it is becoming fixed to consume money through display screens such as smartphones. In addition to this, as the technology related to non-face-to-face is developed, the trend of consumption culture is changing. The unmanned convenience store have also appeared and unmanned cafes are also thriving featuring robot baristas. Cafes operated by robot baristas are even delivering fun to consumers, and hotels and lodging industries are also attracting consumers with unmanned smart check-in and room automation systems. This phenomenon also applies to the financial industry. Online banks such as Kakaobank are growing rapidly. In the fashion industry, untact marketing is being implemented through virtual fitting services that allow users to see their outfits without changing clothes and also through smart stores that allow users to view detailed information such as product color and price without having to ask employees. In this trend, a collaboration between distribution models and finance are also spreading. BGF Retail, which operates a convenience store called CU, has signed an agreement with Shinhan Card to activate unmanned payment services and is working together on payment technology data businesses by biometric authentication payment and cooperating with big data marketing [11]. In particular, the delivery service market is growing rapidly due to COVID-19, and most restaurants are making sales through delivery service applications. With this increase in non-face-to-face, consumers are repeatedly buying and consuming through service applications from service providers such as shopping malls, dining out, delivery and finance [13]. As the changes in the marketing environment, an investigation was conducted on how the empirical value of untact marketing affects consumer satisfaction and consumer's willingness to repurchase, centered on service mobile application. This investigation intends to contribute to the establishment of untact marketing strategy. After using service mobile application that is frequently used in our daily lives, such as delivery apps, shopping apps, and finance apps, consumer satisfaction with their empirical value and their intention to repurchase them was studied and tried to present the direction for the strategic development of untact marketing by revealing the influential relationship. Therefore, empirical values were contemplated, and according to this, concepts of consumer satisfaction, willingness to repurchase were summarized. The research model was constructed based on the theoretical background, and the survey was conducted for empirical analysis. The collected data was obtained using the SPSS 22 program.

2 Theoretical Background and Hypotheses

2.1 Concept of Untact Marketing

The start of untact marketing can be seen as a non-face-to-face service. Non-face-to-face service is a service that customers receive by utilizing information and communication technology without contacting the service provider [12]. With the rapid development of technology, such as the development of the Internet and the spread of social networks, non-face-to-face services have been utilized in various fields and have been subject to research. Internet banking and website service quality research were conducted based on the SERVQUAL scale [8], which is a quality measurement method, and various measurement variables are currently being used to measure the quality of non-face-to-face service [2]. Technology-Based Self-Service (TBSS) is a technical effort by many service companies to deliver services and reduce their workforce in this competitive environment [4]. Recently, untact culture is widely spreading which allows consumers to consume without having to contact employees. Untact is the application of new technology beyond TBSS based technology in terms of adapting to the situation and providing differentiated services. Consumers have become accustomed to and comfortable with this untact technology [5]. In addition, companies actively accept untact marketing to reduce costs caused by the rise in the minimum wage, and consumers can also feel satisfied through unrestricted economic activities of time and space. The information gained through untact marketing attracts consumers into untact marketing [5].

2.2 Empirical Value

Empirical value refers to the benefit or exchange value perceived in the process of objectively evaluating a product or using a service. current empirical value is the creation of feelings, thoughts, actions, and relationships as a way to maximize and streamline the consumer's experience, which is a key part of marketing [7]. Holbrook (1994) defined empirical value as the preference expressed by the activity of using a product or service. In addition, in accordance with intrinsic value, extrinsic value, active value, and reactive value, it was seen as the four factors of enjoyment, investment profitability, service excellence, and aesthetics [3]. Based on the value type of Holbrook (1994), Mathwick et al. [10] developed in more detail. In other words, it was divided into consumer utility, service excellence, enjoyment value, and aesthetic value [9]. In this study, empirical values were embodied and set up as five factors: hedonic value, practical value, economic value, service value and reliability value.

2.3 Consumer Satisfaction Level

In order for companies to satisfy consumers, they need to improve their products and services so that customers can feel satisfied. Therefore, the quality of products and services is an important factor in the company's revenue and customer satisfaction with its customers. Consumer satisfaction was defined as an emotional response, cognitive states, evaluation, and satisfaction judgment. That is, it can be divided into the results of the experience gained through consumption or the consumer assessment of the consumption experience process and it can be divided whether the process or outcome of the consumption experience is an emotional or cognitive judgment [10]. Studies of consumer satisfaction define the concept of consumer satisfaction as a combination of cognitive and emotional responses, while on the other hand as a result from consumption experience, it defines consumer satisfaction as being perceived as being adequately or improperly compensated for the price paid by the consumer [14]. In this study, based on the prior research, the measurement items were composed of benefits, meet expectations, and satisfaction with use.

2.4 Repurchase Intention

The repurchase intention is likely to occur as a real action against an individual's beliefs or intended behavior and it is the extent that an individual continues to use a particular product or service [1]. The repurchase intention is to maintain a long-term relationship with the customer rather than a one-time relationship with the customer and is closely related to the company's profit-making. Therefore, repurchase intention is meaningful to the company. In the preceding research, which studied consumer satisfaction and willingness to repurchase intention, the focus was on the intention of repurchase of products and services, revisit the store, and the intention to recommend the product to people around them. In this study, items on the repurchase intention were measured as continuous purchase, willingness to recommend the application to the others, and increasing purchase items.

2.5 Research Model and Hypothesis

This study is to investigate how the empirical value of untact marketing influences consumer satisfaction and repurchase intention centered on service application. The variables were defined through the theoretical background and the hypothesis were set as follows:

H1. The empirical value of service apps will affect consumer satisfaction.
H2. The empirical value of the service app will affect the repurchase intention.
H3. The consumer satisfaction of the service app will affect repurchase intention.

The research model of this study is to prove that the empirical value of the use of service apps affects consumer satisfaction and repurchase intention, and that consumer satisfaction affects the repurchase intention (Fig. 1).

In the research model, empirical values perceived by consumers using service apps were defined as five factors: hedonic value, utilitarian value, economic value, service value and reliability value. Consumer satisfaction is the consumer assessment of the overall satisfaction status and was defined as convenience, utilization benefits, meeting expectations, and satisfaction with service apps. The repurchase intention is the intention to use it continuously, and it is defined as the plan for use, recommendation, and continuous order schedule.

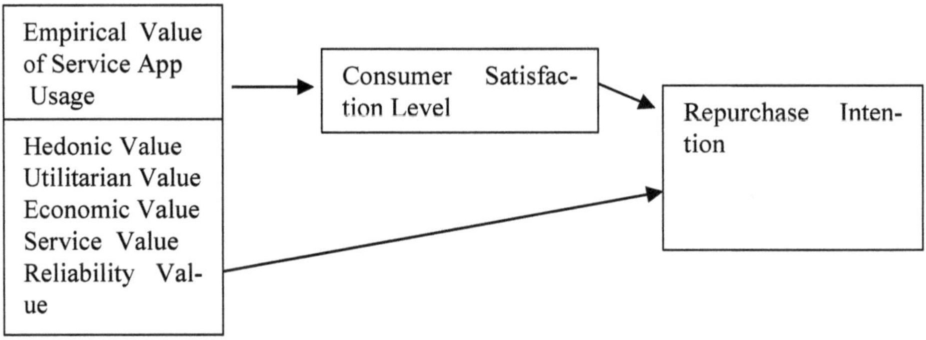

Fig. 1. Research Model

3 Plan for Investigation

The actual investigation were conducted for this study and due to the COVID-19, online survey were conducted. A total of 150 people participated, and the survey were conducted from October 6 to October 18, 2020. The survey configuration consisted of questions about empirical value, consumer satisfaction, willingness to repurchase and demographics. A total of 150 data was used in the analysis, and it was analyzed through descriptive statistic analysis, reliability analysis, factor analysis, regression analysis, and correlation analysis.

4 Empirical Analysis

4.1 Descriptive Statistics and Reliability/Validity Analysis

The demographic characteristics of the sample include 83 males and 67 females, 55.3% and 44.7% respectively, of the total 150 respondents. 71 respondents were aged 10–20 which accounts 47.3%, 72 respondents were aged 30–40 which accounts 48.0%, 7 respondents were aged 50–60 which accounts 4.7%. Of the respondents, there were 52 office workers (34.7%), 38 professionals (25.3%), 33 students (22%), 19 housewives (12.7%), and 8 self-employed (5.3%). 50 respondents were high school graduates or below (33.3%), 49 respondents were college graduates (32.7%), 23 respondents were university graduates (15.3%), and 13 respondents were graduate school graduates (18.7%). Among the respondents, 53 respondents were reside in Seoul (35.3%), 46 respondents were reside in Gyeonggi province (30.7%), 21 respondents were reside in Incheon (14%), and 30 respondents were reside in other provinces (20%). For the average monthly consumption expenditure, 39 respondents were below 100,000 won (26%), 51 respondents were in between 100,000–200,000 won (34%), 38 respondents were in between 200,000–300,000 won (25.3%), 14 respondents were in between 300,000–400,000 won (9.3%), and 8 respondents were 400,000 won or above (5.4%). In reliability analysis, the Cronbach's coefficient alpha was .923 for 21 items including empirical value, consumer satisfaction, and repurchase intention, which shows internal consistency. Factor analysis was done for validity analysis and the results are as shown in Table 1.

Table 1. Credibility/Consistency

Criteria		Communality	Cronbach's α	KMO	χ^2	p-value
Empirical value	Hedonic value	.786	.897	.803	373.066	.000
	Utilitarian value	.741	.644	.500	39.140	.000
	Economic value	.811	.766	.500	72.065	.000
	Service value	.736	.597	.500	37.143	.000
	Reliability value	.671	.849	.795	326.066	.000
Satisfaction level		.788	.911	.855	4284.143	.000
Repurchase intention		.746	.885	.729	470.094	.000

If the KMO, an index that tests the suitability of a factor analysis, is between 0.5 and 1, then it is considered appropriate. As shown in Table 1, the communality is 0.5 or more indicating that the measurement questions are turned out as appropriate factors.

4.2 Correlation Analysis Between Variables

The most commonly used Pearson product-moment correlation coefficient was used to identify the relevancy between variables. The correlation coefficient has a value between −1 and +1. The greater the correlation in the direction of positive (+) or negative (–), it is closer to +1 or −1. In general, if the absolute value of the correlation coefficient is 0.4 or less, it can be said that there is a weak correlation; if the correlation is 0.6 or higher, there is a strong correlation. There is a significant correlation between the factors as shown in Table 2.

4.3 Hypothesis Verification

H1. The empirical value of service apps will affect consumer satisfaction.

For the verification of hypothesis 1, multiple regression was performed. Empirical values were set as independent variables and consumer satisfaction level was set as dependent variables. As shown in Table 3, the value for the R square is .162 with an explanatory power of 16.2%, and the F-value is 5.573 with a significance probability of .000 of a regression equation, which shows significant results. Among the factors of empirical value, hedonic value and utilitarian value were shown to have a positive (+) influence on satisfaction level, however it was shown that economic value, service value, and reliability value does not affect satisfaction level.

Table 2. Correlation analysis between variables

Variables	Hedonic value	Utilitarian value	Economic value	Service value	Reliability value	Satisfaction level	Repurchase intention
Hedonic value	1						
Utilitarian value	.823**	1					
Economic value	.810**	.676**	1				
Service value	.780**	.673**	.753**	1			
Reliability value	.842**	.724**	.765**	.707**	1		
Satisfaction level	−.191*	−.322**	−.250**	−.268**	−.229**	1	
Repurchase intention	.820**	.731**	.744**	.704**	.737**	−.186*	1

** correlation coefficient is valid at 0.01 level
* correlation coefficient is valid at 0.05 level

H2. The empirical value of the service app will affect the repurchase intention.

For the verification of hypothesis 2, multiple regression was performed. Empirical values were set as independent variables and repurchase intention was set as dependent variables.

As shown in Table 4, the value for the R square is .696 with an explanatory power of 69.6%, and the F-value is 66.008 with a significance probability of .000 of a regression equation, which shows significant results. Among the factors of empirical value, hedonic value was shown to have a positive (+) influence on repurchase intention and it was shown that utilitarian value, economic value, and service value does not affect satisfaction level.

H3. The consumer satisfaction of the service app will affect repurchase intention.

For the verification of hypothesis 3, simple regression was performed. Consumer satisfaction levels were set as independent variables and repurchase intention was set as a dependent variable. As shown in Table 4, the value for the R square is .714 with an explanatory power of 71.4%, and the F-value is 45.660 with a significance probability of .000 of a regression equation, which shows significant results. Consumer satisfaction was found to have a positive (+) effect on repurchase intentions. It was shown that the higher the consumer satisfaction, the repurchase intention was higher (Table 5).

Table 3. Effect of empirical value of service apps on consumer satisfaction levels

	Unstandardized beta coefficient		Standardized beta coefficient	t-value	p-value	Collinearity	
	B	Standard error	β			Tolerance	VIF
Constant	3.042	.225		13.497	.000*		
Hedonic value	.602	.202	.581	2.976	.003*	.153	6.557
Utilitarian value	−.511	.145	−.476	−3.516	.001*	.318	3.145
Economic value	−.153	.128	−.169	−1.192	.235	.290	3.447
Service value	−.203	.129	−.204	−1.570	.119	.345	2.900
Reliability value	−.102	.150	−.100	−.681	.497	.268	3.735

R^2 .162 R^2adj .133 F = 5.573 p-value .000 *p < 0.05.

Table 4. Effect of empirical value of service apps on the repurchase intention

	Unstandardized beta coefficient		Standardized beta coefficient	t-value	p-value	Collinearity	
	B	Standard error	β			Tolerance	VIF
constant	−.163	.160		−1.018	.310		
Hedonic Value	.663	.143	.544	4.628	.000*	.153	6.557
Utilitarian Value	−.088	.103	−.070	−.855	.394	.318	3.145
Economic Value	.202	.091	.189	2.221	.028	.290	3.447
Service Value	.126	.091	.108	1.383	.169	.345	2.900
Reliability Value	.129	.160	.108	1.221	.224	.268	3.735

R^2 .696 R^2adj .686 F = 66.008 p-value .000 *p < 0.05

Table 5. Effect of consumer satisfaction level of service apps on the repurchase intention

Dependent variable	Independent variable	Unstandardized beta coefficient		Standardized beta coefficient	t-value	p-value
		B	standard error	β		
Repurchase intention	constant	.093	.076		1.225	.221
	Consumer satisfaction level	.867	.024	.903	36.450	.000

R^2 .714 R^2adj .499 F = 45.660 p-value .000 *$p < 0.05$

5 Conclusions

To summarize the findings, first, empirical values affect consumer satisfaction. It turned out that hedonic value, practical value, economic value, reliable value, and service value which are relatively meaningful outcomes, affect consumer satisfaction as well. Second, empirical values affect the repurchase intention and hedonic value is shown to have a positive (+) effect whereas utilitarian, economic, service, reliability value affect less. Third, the consumer satisfaction level has been shown to affect the repurchase intention. In this study, consumers seem to think positively about untact marketing, and as the consumer market for the untact marketing expands, it is believed that corporate strategic efforts are needed so that consumers can feel interested, as well as realizing economic, and rational value. For future studies, it will be meaningful to study the relationship between consumer satisfaction of detailed methods of untact marketing and repurchase intention to examine what is effective untact marketing.

References

1. Bhattacherjee, A.: Understanding information systems continuance: an expectation confirmation model A no. 1. MIS Q. **25**(3), 351–370 (2001)
2. Choi, H.Y.: A Study on the Structural Relationship between Service Channel Usability and Customer Relationship Quality and Loyalty of Airline, Doctoral thesis, Graduate school of Kyung Hee University, vol. 11 (2011)
3. Holbrook, M.B.: The nature of customer value: an axiology of service sin the consumption experience. In: Rust, R.T., Oliver, R.L. (eds.) Service Quality: New Directions in Theory and Practice (1994)
4. Kim, I.H.: The Influence of TDOO Characteristics and Personalities on Intention of Use, Master's degree thesis, Graduate School of Chung-Ang University, vol. 22 (2010)
5. Kim, N.D.: Trend Korea 2018, MiraeueChang, pp. 313–334(2018)
6. Kim, Y.T.: Go for untact marketing: changing consumption culture. Monthly CEO **128**(5), 55–57 (2020)
7. Ko, G.S.: Experience Marketing. Daemyeong Publishing, Daegu, vol. 25 (2004)
8. Li, Y.N., Tan,K.C., Xie.,M.: Measuring Web-based service quality. Total Qual. Manag. **13**(5), 685–700 (2002)

9. Mathwick, C., Naresh, M., Edward, R.: Experiential value: conceptualization, measurement, and application in the catalog and internet shopping environment. J. Retail. **77**(1), 39–45 (2001)
10. Park, M.H., Cho, H.J.: Reestablishment of customer satisfaction. Asia Mark. J. **1**(4), 126–151 (1999)
11. Park, S.H.: Spread untact marketing. Maekyeong Econ. **2001**, 68–69 (2019)
12. Seo, C.J., Kim, J.H., Lee, S.Y.: A study on the measurement model of non-face-to-face service quality. J. Korea Serv. Manag. Soc. **8**(1), 111–135 (2007)
13. Shin, S.Y.: Startup-driven Delivery App Market. http://www.m-i.kr/news/articleView.html?idxno=517586
14. Yi, Y.J.: A critical review of consumer satisfaction. In: Zeithaml, V.A. (ed.) Review of Marketing. American Marketing Association, Chicago, IL, pp. 68–123 (1990)

A Study on the Influence on Intention to Use Blockchain-Based Copyright Contract

Focusing on Blockchain-Based Digital Copyright Exchange Core Platform

Jung Jae Lee(✉)

Department of Entertainment and Art Management, Soongsil Cyber University, Seoul, Korea
jjlee@mail.kcu.ac

Abstract. The creation, distribution, consumption, distribution and settlement of works in the development of information and communication brought about a paradigm shift across the ecosystem of works. The consumption pattern of owning and consuming content has changed into a form of sharing, and the life cycle of content in the consumption market of works is developing very rapidly, and it is gaining popularity around the world, including K-pop and dramas, amid the Korean Wave.

In the case of digital content copyright protection and safe transactions, there are frequent cases where rights cannot be claimed due to the failure to exceed the walls of efforts and costs spent on utilizing trusted institutions and means, therefore, a platform that enables secure rights to the work and safe transactions between individuals is essential by introducing block chain technology, and a method of securing trust in the distribution of digital content, integrity and copyrights by using the integrity of transactions of the block chain technology.

This study finds and utilizes stakeholders' active involvement in the activation of digital copyright exchanges through the development of block chain-based copyright contracts and distribution platform technologies that provide a total service from copyright contracts to distribution by developing high-capacity content data storage and retrieval technologies, high-speed real-time transaction processing technologies, and content distribution technologies, which are core block chain technologies that support the distribution of digital content among creators and producers based on transparent and reliable block chain-based smart copyright contracts.

Keywords: Content · Content distribution · Copyright · Blockchain · Contract

1 Introduction

The size of the digital content market continues to grow, with sales of the domestic content industry reaching 113.216 trillion won and value added 44.229.8 trillion won in 2017, with the value-added ratio reaching 39.1%. In addition, exports amounted to $8.814.44 billion in 2017, while imports amounted to $1.2436 billion, showing a trade surplus of $7.61 billion and the number of businesses also stood at 105,475. In terms of

R. Lee (Ed.): ICIS 2021, SCI 1003, pp. 96–106, 2022.
https://doi.org/10.1007/978-3-030-90528-6_9

boosting the content industry, sales from 2018 to 2018 have grown 5.4% over 18 years to 122.6 trillion won, while exports have grown 8.5% to 81.3 billion dollars. As such, the national content industry continues to grow, playing a pivotal role in the Korean Wave, and in the past, game-oriented limited content exports have recently shown remarkable achievements in all areas, including K-pop, dramas, movies, characters, cartoons and publishing [9].

In the content sector, mobile-based content has been in the spotlight for many years, and sales in the domestic mobile content industry in 2018 were estimated at 23.3 trillion won, and it is expected to increase 11.7% annually from 2016 to about 31.52 trillion won in 2021 [9].

While the mobile game sector, which has the largest sales volume, maintains solid growth at an annual average rate of 8.7% until 2021, most sectors except the mobile communication and entertainment-music sectors are expected to continue to grow at an annual average rate of more than 10%.

Recently, global content platforms have been established due to the development of devices and networks, and new content markets such as 5G-based virtual reality in 2019 and various cloud-based services are creating a new market environment.

The Digital Copyright Exchange is being established and operated to cope with changes in the copyright environment and to provide copyright information in an integrated manner for the registration, change tracking, security, search, report and analysis of copyright information and increase the efficiency of copyright work.

In terms of platform for distribution of works, data on works are registered in integrated repositories after standardization process is carried out according to standards, and users of works can receive various standardized information from integrated repositories and use it for services, and new service providers can easily enter the business through initial cost savings and time savings, and provide opportunities to generate new revenue [1].

The digital exchange consists of active participation by stakeholders such as rights holders, distributors, and service businesses by playing a role in supporting transactions between rights holders and service providers and users in a convenient and easy way. However, there are situations in which the efforts and costs of utilizing trusted institutions and means to protect the rights of digital content works and secure transactions are not exceeded, and a platform is needed to enable secure transactions between copyright holders and individuals (or between organizations and businesses) at no extra cost.

As a result, Blockchain technology using transaction details integrity is receiving attention recently, and technical alternatives such as controlling data access rights for reliable content distribution, efficient storage of exponentially increasing data, and search functions for distribution activation are needed.

In this study, to support the transparent, reliable, smart environment-based copyright contract between copyright holders and users, we would like to apply the Blockchain to analyze the validity of the application and utilization at the contract stage.

The research method derives requirements and evaluation criteria for establishing a Blockchain-based contract model through research on trends on the development of digital content copyright contract and distribution technology based on the Blockchain, and gathering expert opinions, and carries out the elicitation and analysis stages for the necessary technology.

In addition, we will develop technical algorithms such as the establishment of the Blockchain-based copyright smart scheme logic, and establish development plans based on simulation performance and derived results for analysis of the validity of algorithms.

Through this, we will establish a low-cost and high-reliability environment between copyright holders and users, and study the applicability of large-capacity content contracts to confirm the value of the copyright industry as a core technology that can raise the technological competitiveness of the copyright industry to the next level.

2 Theoretical Background and Hypotheses

2.1 Digital Copyright Exchange

The "Digital Copyright Exchange" is a new concept created by the Ministry of Culture, Sports and Tourism and the Korea Copyright Commission, and since 2008, actual work has been carried out mainly on permission contracts. It includes the concepts of digital, copyright and distribution (exchange) as key keywords in the content industry [5].

Copyright rights groups and operators handle the creation of copyright and copyright information, distribution (use and transaction), and distribution of settlement, but managing properly used details in numerous copyrighted places requires a system that can ensure transparency and reliability in a series of processes. In addition, in the event of a copyright dispute, the value of copyright should be calculated along with the appraisal and adjustment of copyright, but the lack of criteria for judging the value of copyright does not activate copyright transactions.

In order to overcome these pending issues, the Digital Copyright Exchange of public character was established for the purpose of legal and cheaply providing digital contents to users, transparently providing revenue to rights holders, and for the purpose of operation and management.

The Digital Copyright Exchange aims to provide copyright information in an integrated manner for the registration, change tracking, security, search, report and analysis of copyright information and to increase the efficiency of copyright work.

From a functional perspective, the data on the work is first standardized in accordance with the standards and then registered in an integrated repository, while the users of the work (distributor, service, organization, etc.) receive and provide services with a variety of standardized media content from the integrated repository. By providing standardized technology and rule-based data to new customers, services are immediately available and opportunities to generate new revenue through expansion into the global business.

It is also involved in the distribution of illegal contents, plays a role in supporting the promotion of distribution through joint legal action, marketing and promotion support for domestic and overseas distribution, and manages the lifecycle of the creation, distribution, use, and extinction of works, enabling the establishment of a sound order of use and the expansion of the distribution base of the copyright industry through the activation of the use.

2.2 Copyright Integrated Management System (CIMS)

The CIMS (Copyright Integration Management System) is a system established for the purpose of registering and managing the Integrated Copyright Number (ICN) and Copyright Information (Meta DB) given to works [4].

UCI refers to a unique string or number system for verifying rights management information of copyrighted works, and is a unique identification number for copyrighted works used by CLMS, which allows the common information such as the real name, stage name, and contact information to be automatically linked when joining a copyright trust organization [3].

Meta DB is information that is stored in the form of a database so that information about works, authors, actors, etc. can be checked at once in connection with UCI, which is a unique identification number [2]. Meta DB has different information needed for each copyright field, and Meta DB is also held by related trust organizations, so trust organizations that own Meta DB must be linked to CIMS in the copyright trading system.

2.3 Blockchain Overview

Blockchain was first introduced in 2008 by Satoshi Nakamoto in the course of developing Bitcoin and means connecting blocks, which mean transaction information data, to chains created based on P2P (Wiki Baggwa 2018). Blockchain is a data forgery prevention technology that cannot be modified because it technically encrypts and stores information such as contracts and transactions on distributed ledger [8].

The definition of a Blockchain may differ from researcher to researcher as shown in the following table, but it contains a common definition of a technology that distributes and stores transactions to all members based on reliability, stability and efficiency (Table 1).

2.4 Blockchain Structure

A Blockchain is a block containing transaction information that is linked to a block containing prior transaction information, which consists of the part, block size, and header in which transaction information is recorded (Antonopoulos 2015).

Transactions in the Blockchain-based service are made by all participants in the Blockchain to share the block containing the transaction history, and when a new transaction occurs, the block containing the transaction record is created, and the information in the new block is sent to all participants to indicate the process in which the validity of the transaction is made. Afterwards, the connection of the block is determined by the approval of the participants.

2.5 Digital Contents Market

The global contents market rose 5.5% on-year to $2.81 trillion in 2016 as the advertising and knowledge and information markets, which control about 48% of the total market, showed growth of more than 6%. The global content market is expected to grow 4.2% annually over the next five years to reach 2.561 trillion dollars, increasing from offline content to online content, and the global market size and outlook for each content are as shown in Table 2 below [10].

Table 1. Define blockchain

Researcher	Definition
Collins (2012)	Transaction database connected to multiple computers (nodes) using bitcoin crypto currency
Oxford Dictionaries (2016)	Digital books in which transactions of bitcoins or other crypto currencies are recorded sequentially and publicly
Jungsuck. Kim&Kwangyong. Kim (2017)	A form of distributed database in which the structure for storing and processing data consists of the connection of blocks
Meeyepn. Kim&Heungyeul. Yeum (2017)	Block containing transaction details is connected in chronological order to form a chain. Open distributed ledger in which block-chain participants, not central managers, create blocks
Jungsook. Kim (2017)	Distributed computing technology that provides reliability, stability, efficiency, and security
Kijin. Chan (2017)	A technology that not only records financial and non-financial transaction data, but also sends and receives transaction information between users, verifies accuracy and reliability, and records financial and non-financial transaction data in an unmodifiable form
Science and Technology Policy Institute (2017)	Means a digital ledger in which transaction information occurring in a public or private network is encrypted and shared among the members of the network

2.6 Copyright Market

According to the current status of copyright registration/management managed by a domestic trust management organization, 25.21 million in 2013, 27.36 million in 2014, and 32.72 million in 2015, the total amount of copyright fees collected in Korea increased by about 32.4 billion won in 2015, and is increasing by 43% compared to 2011.

In 2015, the collection of user fees for each field of work was 166.4 billion won (74.1%), followed by 24.5 billion won (10.9%), and 23.4 billion won (10.4%), with music accounting for 74% of the total. Considering that the number of managed works in the music sector is 21.3% of the total, the total amount of collection is very different at 74.1% of the total, while the total amount of collection in the news sector, which accounted for 68.3% of the total number of managed works, is only 4.4%.

Table 2. Digital contents market

Type	(Year of 2016)	(Year of 2017)	(Year of 2018)	(Year of 2019)
Publication	332	330	328	325
Comics	7	7	7	7
Music	47	49	51	53
Game	94	109	124	135
Movies	40	41	43	45
Animation	7	8	8	8
Broadcasting	488	500	513	526
Advertisement	517	547	574	597
Knowledge information	667	708	748	787
Character	263	277	290	302
SUM	2,462	2,575	2,687	2,786

Source: Korea Creative Content Agency 2017 Overseas Content Market Trend Survey

3 Research Method

3.1 Purpose of Research

The purpose of the research is to promote the distribution of the Blockchain-based digital contents, which provides integrated services for copyright contracts. The goal is to establish a safe copyright distribution environment without fostering reliability of digital copyright exchanges through research on large-scale content data storage and retrieval technology, ultra-fast real-time transaction processing technology, and content copyright contracts based on content identification system, which are core Blockchain technologies that support digital content distribution.

3.2 Overview of Research Activities

Blockchain-based copyright smart contract technology consists of smart contract-based contract model research, smart contract-based copyright contract optimization technology research, consumer-centered content matching technology and algorithm research, Blockchain settlement distribution, profit sharing model design, Blockchain-based contract platform support technology research, and digital content copyright identification system research, and each subject is as shown in Table 3 [7].

3.3 Research Promotion Stage

Research activities aimed at developing new digital copyright exchange services and products through the steps of research planning, necessary technology derivation and analysis, design and construction, integration test, system test, post review and evaluation, and commercialization.

Table 3. Blockchain-based copyright smart contract technology study

Research item	Research content
Smart contract-based contract model study	Research on contract recommendation technology by contract data type Study on the smart contract contents prevention technology
Smart contract-based copyright contract optimization technology study	A Study on the Optimization of Smart Contracts for Digital Contents, such as Video/Music/Books Study on the automatic generation technology of smart contract-based standard contract
A study on the consumer-oriented content matching technology and algorithm	Derives needs for copyright holders and distributors using FGI Model FGI results and derive content contract algorithm
Blockchain settlement distribution, profit sharing model design	Research on technology for linking log information of content works based on Blockchain for settlement/distribution A study on the design of revenue sharing for the entire process of contract signing (LifeCycle) based on Blockchain
Blockchain-based contract platform support (common) technology study	A study on content copyright registration/verification/distribution/settlement management plans through distributed ledger
A Study on the content copyright identification system	Content copyright identification system structure and standardization model design Content copyright identification system development

In the first research planning stage, research and analysis of trends in the development of Blockchain-based digital content copyright contracts and distribution technologies, and technology development and systematization according to the results are essential, and a trend analysis report for major contents is constructed along with the survey [9].

The second step is to clarify the necessary skills and procedures for development, plan for the work on the entire task and quarterly tasks, and prepare a task table, to derive the necessary skills, analyze the technical derivation matters, create a task table, Perform the quarterly task schedule in order.

Third, in the design stage, the structure and detailed design are performed, and the related tasks are developed, work plans are established, and the development schedule is managed. In this stage, a Blockchain-based copyright smart contract technology was designed and built to develop a technology related to a Blockchain-based digital content copyright contract [6].

Fourth, in the integrated test stage, each implemented research project is integrated into one to verify that it operates as designed, and in the system test stage, the actual system environment is configured to install the integrated research project and verify that the function is performed as required do.

Lastly, post-review at the end of each quarter, the core technology factor (CTE) demonstration, and then the measurement of the results of the quarter, all research processes are completed.

3.4 Core Technology

Core technologies include: ① Blockchain settlement distribution, revenue sharing technology, ② Blockchain-based contract platform support-based (common) technology, ③ Ultra-high-speed real-time transaction processing technology development, ④ Content identification system-based content distribution technology development, ⑤ Large data distribution Blockchain Platform API and Toolkit development, ⑥ Blockchain-based copyright smart contract and distribution platform construction.

4 Empirical Analysis

4.1 Contract Model Design Using Blockchain

Direct transactions between users and copyright holders without the intervention of a trust organization or proxy brokerage, and aims to freely trade as a user and copyright holder, and establishes a Blockchain-based copyright prosumer model for completion of registration/distribution through a smart contract of the Blockchain (Fig. 1).

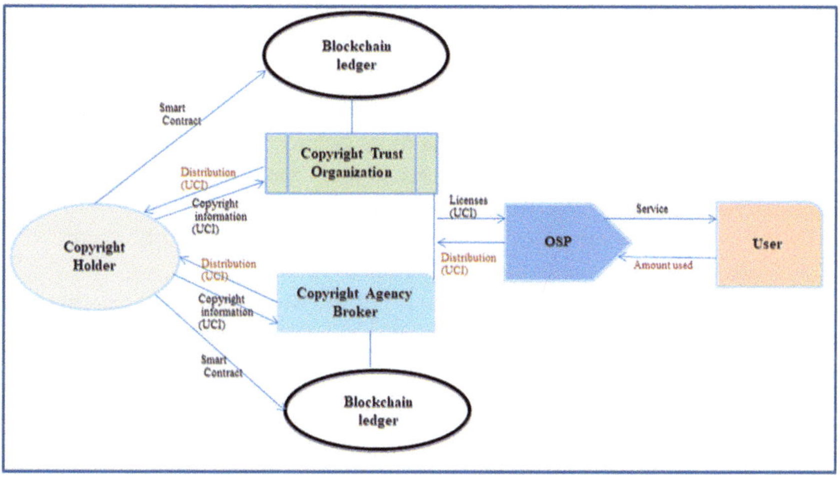

Fig. 1. Contract model using blockchain

To build a Blockchain-based OSP-linked work model, copyright information and usage contract information are created as a smart contract and distributed to the

Blockchain, and the copyright holder creates protected content using encryption (DRM) technology linked to the smart contract, and Blockchain-based OSP). At this time, the main function is the registration and verification of ownership/copyright through the distributed ledger, and the OSP of the work is a service, but the authority to access/use the work is performed based on the Blockchain smart contract.

In the Blockchain-based copyright trust company and proxy intermediary linkage model, the copyright holder registers a smart contract that includes copyright information and a license and distribution rule on the Blockchain. At this time, the trust organization provides the copyright information and license information registered on the Blockchain ledger to the OSP business operator, and the OSP business operator remits the proceeds to the trust organization through the Blockchain. The trust organization automatically pays the copyright holders and performers according to the distribution rules of the smart contract.

4.2 Blockchain Contract Content Exposure Prevention Technology Research

Blockchain basically guarantees data integrity, but does not provide confidentiality services, so it implements sophisticated cryptographic technology to prevent the disclosure of some transaction data or allow access only to authorized persons.

At this time, in the case of a private or consortium Blockchain, it provides a confidentiality service that provides user authentication and access control services at the participation stage.

4.3 Blockchain Contract-Based Copyright Contract Optimization Technology Research

Blockchain contract optimization specialized for digital content works such as video, music, and books, and smart contract-based standard contract automatic generation technology were developed, and the research contents of these technologies are as follows (Table 4).

In addition, if Blockchain technology is to be operated in connection with existing copyrighted works and distribution systems, technical measures necessary for interoperability such as data conversion and connection protocol must be prepared. Based on the UCI code system, it is necessary to design the data standard to be recorded in the Blockchain ledger and to migrate the existing data.

It is necessary to design a protocol that allows the Blockchain and existing systems to communicate with each other and exchange data. New protocols can be designed, and existing SOAP, RPC web-based protocols can be used. These tasks are designed in consideration of security as well as ease of maintenance and scalability of the linked system.

Table 4. Research content of smart contract-based copyright contract optimization technology-copyright contract optimization technology

Research item	Research content
Research on optimization of smart contracts specialized for digital content works such as video, music, and books	Correction of copyright information, copyright information, etc. Prevention of excessive computing resource consumption due to mining competition Blockchain division preparation technology Confidential service provision technology Standardization of data and business processes of participating organizations Blockchain transaction processing cycle by business type Blockchain security measures Linkage with existing digital copyright exchange CLMS
Smart contract-based standard contract automatic generation technology	Smart contract is a 'automated contract' system design that automatically executes the contract when all programmed conditions are satisfied Smart contract designates the contract terms in computer code and executes the contract when the terms are met Through smart contracts, people can trade various things such as copyright, settlement, and usage fees, and implement technology that enables transactions between parties without a third party

5 Conclusions

In the shared ledger of the Blockchain, information protection, sharing, and transparency contradict each other, but through the development of encryption technology and access control technology suitable for the shared environment of the Blockchain, the technical limitations of the Blockchain could be overcome. In particular, the existing Blockchain technology overcomes the technical limitation that content cannot be distributed due to capacity constraints, thereby maximizing the utility of the technology, and securing the base technology to be utilized and grafted in various copyright transaction environments.

In addition, this research can act as a core infrastructure technology for content exchange and distribution, and in a situation that is recognized as a global cultural content powerhouse such as K-Pop and K-Drama, it has an indirect ripple effect on the technological competitiveness of the content industry, expectations became possible.

Through the results of this study, it is possible to develop various business and service models by linking with the existing digital copyright exchange CLMS, and it can evolve into a variety of data trading platforms. It is expected to be able to establish itself as a core technology for an unlimited content distribution platform.

References

1. Korea Software Promotion Agency: A study on establishment of digital content distribution framework and technology standard strategy (2003)
2. Jeong-hoe, H.: Research on identification metadata of national URN-based identification system, Korea Information Society Agency (2003)
3. Trend Report: Research on distribution activation plan using UCI identifier, Korea Creative Content Agency (2010)
4. Korea Copyright Commission, Research Report: A study on the status of information systems and linkage plans for efficient use of domestic and foreign copyright information such as ICN (2010)
5. Research Report: Digital Copyright Exchange Operation Strategy Establishment (2011)
6. Kim, M., Yeom, H.: A study on the use cases of Blockchain. J. Korean Inst. Commun. Sci. (2017)
7. Jeong-suk, K.: Analysis of Blockchain-based service status and problems, J. Converg. Knowl. Soc. (2018)
8. Min, Y.: A study on how to use Blockchain technology to activate digital content transactions as a sharing economy. J. Korea Comput. Inf. Soc. (2018)
9. Kim, B.-R.: A Study on the Factors Affecting the Intention to Accept Blockchain-based Music Services, Chung-Ang University Graduate School (2019)
10. Korea Creative Content Agency: Survey on Overseas Content Market Trends (2017)

A Study on the Direction of Beauty Tech Reflecting the Skin Characteristics of Koreans: Focused on Case Studies

Yoo Jeong Lee(✉), Ji Woo Choi, Hyun Woo Nam, and Sae Young Shin

Seogyeong University, Seoul, South Korea
uwjd8882@naver.com

Abstract. The cosmetics industry continues to develop products related to personal curation in convergence with the fourth industry as consumers' needs for personalized solutions increase amid the ever-growing and vast beauty market and information worldwide. In response, the research aims to present a strategic alternative to the development of the smart beauty industry related to personalized beauty products based on big data on skin information tailored to consumer needs. To this end, the examples of beauty technology based on artificial intelligence technology were analyzed by combining literature research and case analysis. Based on this, FGI was conducted and analyzed based on the integrated report and prior research of Koreans to identify the skin characteristics of Koreans. In conclusion, with continuous research and collection of various information data according to skin characteristics, we will be able to provide competitive beauty-tech services and products by increasing the accuracy of the service.

Keywords: Beauty tech · Skin data · 4th Industrial revolution · Smart beauty · Artificial intelligence

1 Introduction

With the arrival of the '4th Industrial Revolution Era' introducing new technologies and leading innovative changes worldwide, modern society is switching into the form of convergence industry, where complex growths of products and services are repeated in most industries [1]. In addition, technologies related to ICBM (IoT, Cloud, Big-data, Mobile) are rising as the core element with the coming of the smart era, leading the next-generation industrial revolution that brings innovative changes through the convergence in overall economic society [2].

Meanwhile, beauty industry is an area of service industry with high purchasing frequency, an easy connection to service, versatility in positioning from necessity to

This work was supported by the Technology Innovation Program (or Industrial Strategic Technology Development Program-Design Innovation Capacity Enhancement Project) (20009851, Development of AI IoT beauty self-care products and distribution services design platform for enhancing global competitiveness of K-Beauty business) funded By the Ministry of Trade, Industry & Energy (MOTIE, Korea).

R. Lee (Ed.): ICIS 2021, SCI 1003, pp. 107–118, 2022.
https://doi.org/10.1007/978-3-030-90528-6_10

luxury item and relatively low entry barrier, and has an outstanding job creation effect; cosmetics industry is expected to make a continuous growth, which will continue to until year 2022 [3]. Accordingly, consumers' needs for personal curation and personalized solutions increased as they face difficulty finding suitable services and products in the midst of extensive beauty industry and information, which resulted from quantitative expansion [4].

In addition, with the technical innovation of smart beauty care in the 4th Industrial Revolution Era, portable skin analyzers, skin analyzer apps, cosmetics curation apps, etc. are continuously developed in addition to personalized cosmetics that focus on individual consumers. Therefore, this study aims to suggest a strategic alternative to the development of smart beauty industry related to personalized beauty products based on skin information that suits consumers' needs. For this, literature review and case analysis are combined to collect cases of beauty tech based on artificial intelligence (AI) technology, and objects are analyzed based on the characteristics of beauty tech. Next, FGI is performed, followed by analyses of integrated reports on Koreans prepared by Foundation of Korea cosmetic Industry Institute from 2011 to 2019 and previous studies to examine the skin characteristics of Koreans. Through this process, this study aims to suggest the direction of beauty tech using skin information big data.

2 Theoretical Background

2.1 Definition and Characteristics of the 4th Industrial Revolution

The 4th Industrial Revolution is the next-generation industrial revolution occurring by the convergence of information and communications technology (ICT) [5]. Starting in the 18th century, Industrial Revolutions have led significant changes in social and economic structures, and continued the greatest advancement in the history of mankind to the 4th significant industrial era since the first industrial revolution. The key of this technical revolution is the combination with new technology through the convergence of areas, such as big data analysis, AI, IoT, mobile, etc., with the existing industry.

It can be characterized as the next-generation industrial revolution where high-tech ICT such as AI, IoT, big data and mobile converges with the economy and society in general to bring innovative changes [6]. Another characteristic is the superintelligence of industrial structure through the connection and convergence of AI and massive big data through advanced capacity, which uses artificial neural network, deep learning, as a medium [7].

The 4th Industrial Revolution became the term that represents the new industrial era of innovative changes through ICT-based convergence, and collapsed the boundary between industries. Also, the 4th Industrial Revolution Era is the innovation of computer-based production method, which can be summarized as the intellectualization of objects through the combination of existing industries that connect product and service through network and new technologies through statistical analysis of big data, as opposed to capital or labor [8].

2.2 Beauty Tech Industry

Beauty tech is a combination of the term 'beauty' and 'technology', and refers to the 'combination of beauty and IT'. As the 4th Industrial Revolution rose as a global issue, IT became a trend in the beauty industry as well, and this trend is expected to have a significant effect on the future new beauty industry paradigm. Beauty tech has a wide range of areas to combine with, including big data, mobile, cloud, Nano technology, AI and IoT, in addition to electronic devices.

Global research firm PNC estimated that the global beauty tech market will make 19.1% of annual average growth and reach 100.9 trillion KRW in market size by 2023. As beauty market grows, the competition between relevant businesses became more intense that consumers options became more diverse, and beauty businesses need to focus on determining what consumers want, and developing core competency. Products combined with beauty tech can especially understand the needs based on skin analysis data and provide skincare and makeup appropriate for the user.

2.3 Skin Analysis and Skin Data

Skin analysis first began as Helene Rubenstein categorized skin type into normal, combination, dry, and sensitive in 1910, and the Baumann skin type, which is the most commonly used now, divides 16 skin types based on 4 indexes (oily (O), sensitive (S), pigmented (P), wrinkled (W)) further into oily (O) or dry (D), sensitive (S) or resistant (R), pigmented (P) or non-pigmented (N), and wrinkled (W) or tight (T) through 65 questions [9]. Physiological property of skin varies generally depending on oil and moisture content [10], and may always change based on age, external environment such as climate and season, sleep, use of cosmetic products, and lifestyle including diet [11]. Skin analysis data is fundamental to beauty tech, and is applied to the skin evaluation model developed to categorize skin type and suggest personalized service, product, or device to individuals. At this time, skin evaluation model is developed based on genetic, environmental, cultural, lifestyle information, etc. therefore, it may show differently by nation, race, and region.

3 Research Method

3.1 Data Collection

For the case study of beauty tech, the case analysis targets were analyzed only as information-provided beauty technology and product beauty techniques among beauty technologies related to skin-tailored and personalization services introduced at CES (Consumer Electronics Show), the world's largest consumer electronics fair. Based on this, six expert Focus Group Interviews (FGI) were conducted on November 7, 2020. After the FGI progress, we tried to analyze the results of prior research after collecting opinions that it is important to set up a model for analyzing skin data based on beauty technology. In addition, prior research and the Korea Cosmetics Industry Research Institute analyzed the skin condition survey report of women and men residing in Korea from 2011 to 2019. The selection of the leading thesis was made through academic research information service (www.riss.kr) and Google search (https://scholar.google.co.kr) to explore the papers searched by keywords 'Skin Type', 'Skin Type Solutions', 'Korean Skin Type' and 'Bauman Skin Type'.

3.2 Measurement

Based on analyzed data, pathfinder method was used to suggest direction for the agenda 'Skin AI-based Beauty Tech Skin Evaluation Model Establishment in K-Beauty and Commercialization of Contents' in the FGI (Focus Group Interview) held on November 7, 2020. 6 experts (1 professor in AI, 1 expert in K-wave contents, 1 professor in content development, 2 professors in skincare, 1 professor in anatomy and physiology) participated in FGI, and the main agendas were as follows: First, setting the approach and direction for building skin artificial intelligence learning DB, Second, expanding the K-beauty market and identifying user requirements for beauty technology, Third, big data construction and research for skin assessment model establishment were divided into cautions, and the conclusions drawn were presented in 4.2 of this study.

4 Result

4.1 Beauty Tech Case Analysis

According to the report by start-up accelerator, 'Future Play', these beauty tech businesses can be categorized into 'information delivery', 'channel', 'product', and 'beauty service'. Among these categories, types of beauty tech were limited to 'information delivery' and 'product' to perform case analysis (Table 1).

Table 1. My skin track series

Product name	My skin track UV [12]	My skin track pore [13]	My skin track pH
Product image	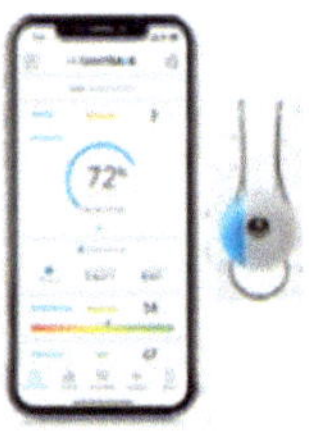	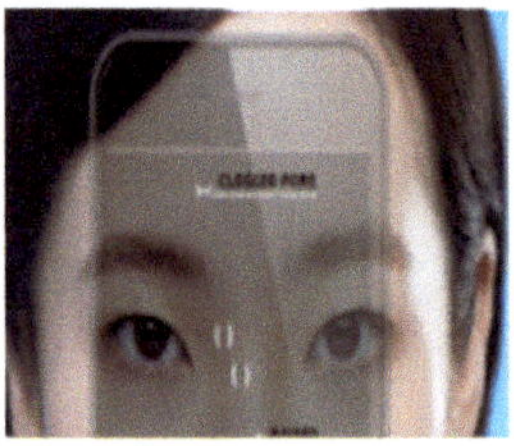	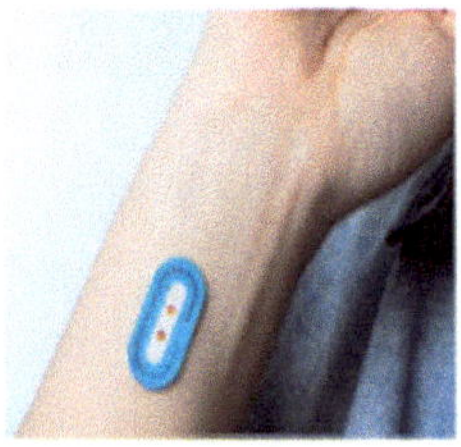
Interface	Clip-type wearable device & smart device app	Smart device, PC & smart device app	Patch-type wearable sensor & smart device app
Beauty tech type	Information delivery beauty tech		

(continued)

Table 1. (*continued*)

Product name	My skin track UV [12]	My skin track pore [13]	My skin track pH
Data information	UV, air pollution level, pollen, temperature · humidity	Clogged pores, raised imperfections, residual marks	pH
Method	Measures UV, pollution level, etc. based on outdoor exposure through the device in real time and alerts at risk level	Takes photos of front, left · right sides and analyzes	Captures sweat through wearable sensor attached on skin and measures pH level
Service	Provides personalized information and suggests customized cosmetic products after analyzing data		

'My Skin Track' series by La Roche Posay measures various environments and exposure to harmful outdoor environments such as UV, pollutants, pollen, heat, humidity, etc. in real time to received personalized advice according to individual setting on skin type, skin tone and interest. It also analyzes visible skin concerns such as pores to determine skin needs, and measures pH through a trace amount of sweat to provide feedback on the result. Like this, collection and analysis personal information including skin information, age, location, and big data-based information data including search keyword, temperature, humidity, air pollution, etc. show that this is an example of 'information delivery' beauty tech (Table 2).

Table 2. Skin measuring instrument

Product name	LUMINI [14]	Skin 360 [15]
Product Image	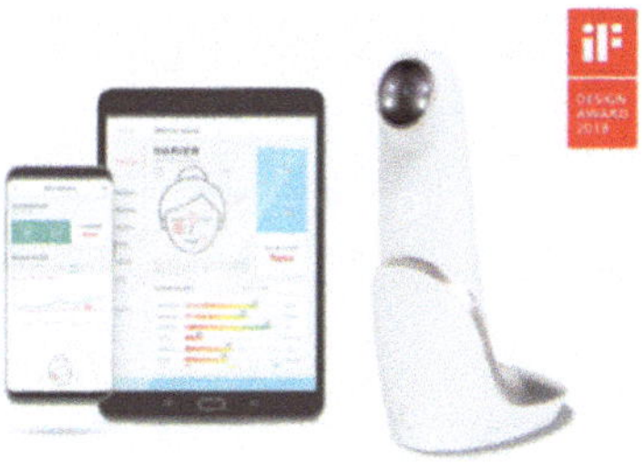	
Interface	Device-linked smart device app	Mobile beauty device and linked smart device app

(*continued*)

Table 2. (*continued*)

Product name	LUMINI [14]	Skin 360 [15]
Beauty tech type	Information delivery, product beauty tech	
Data information	Skin information- wrinkle, pigmentation, redness, pores, sebum, skin trouble, etc	Skin information- dark circle, wrinkle, fine wrinkle, smoothness, dark spot, glow, moisture, even tone, clear skin
Method	Face photo (front) – Skin analysis through deep learning – Product, procedure recommendation through big data analysis	App analyzes selfie to evaluate skin health and predict progress based on time progress
Service	Analyzes data, and provides personalized information and customized cosmetic products for skin improvement	Recommends skincare products, measures individual skin score

Both 'LUMINI' and 'Skin 360' display the characteristics of information delivery beauty tech, which links smart device app and the device to measure skin and analyze skin data such as wrinkle, pigmentation, dark circle, skin trouble, etc. These apps are built using the database of over 10,000 actual users of all ages, gender, and races in order to understand different skin types. in addition, it integrates real-time weather data to include external factor. at the same time, they recommend personalized product or cosmetics based on such information, which shows that they display the nature of product beauty tech as well. Like this, both products have properties of information delivery beauty tech and product beauty tech, so users can acknowledge their skin issues and solve the issues based on recommendations of products that these products suggests as solutions (Table 3).

Table 3. Mask pack 3D printer

Product name	MaskiD [16]	Amore 3D printing pack [17]
Product image		
Interface	Mobile app-smart device link, 3D printer	
Beauty tech type	Product beauty tech	

(*continued*)

Table 3. (*continued*)

Product name	MaskiD [16]	Amore 3D printing pack [17]
Data Information	Facial area such as forehead, eye area, cheeks, nose, columella, etc.	Position of eyes, nose, lips, area data of forehead, cheeks, chin, etc., skin condition
Method	Creates multidimensional map based on facial condition information, and creates mask pack with 3D printer	3D printer prints with facial and skin condition data obtained from facial image captured using mobile app, and provides customized cosmetic product
Service	3D Printed mask pack	

Both MaskiD and Amore 3D Printing Pack were developed to solve the problem of existing mask packs, which could not reflect different face sizes, location of eyes, nose and mouth, skin color and skin defects by users. They analyze individual face and combine AI and 3D printing technique to design and provide personalized mask pack. 3D printer prints out a personalized mask pack for the user based on the information of position or shape of face. Both products have the properties of product beauty tech as they create products based on big data (Table 4).

Table 4. Custom formla [18]

Product name	Perso	Opte precision skincare
Product image		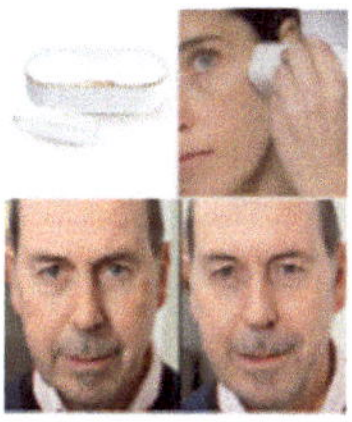
Interface	Smart device & device	Device
Beauty tech type	Information delivery, product beauty tech	Product beauty tech
Data information	Skin information data, environment information, etc.	Skin information such as skin tone, etc.
Method	Skin analysis-Surrounding environment verification-Selection to create customized formula	Device scans face and covers blemishes, etc.
Service	Creates customized formula	Smooth skin expression

Both devices recommend products and cover blemishes based on the skin measurement, displaying the property of product beauty tech, but perso also analyzes skin and checks the surrounding environment to provide skin-related information to the user, which is a property of information delivery beauty tech. Both products are device-based, and in addition, perso measures skin through the app on smart device. perso also demonstrates L'Oreal's goal to establish a new ecosystem based on beauty tech, which conveniently enables refill and purchases through NFC attached on the cartridge. P&G's Opte Precision Skincare performs precision skincare according to the skin tone. more specifically, it scans the face and accurately camouflages any spots discovered for smooth-looking complexion. In other words, it is like a smart inkjet printer for the face. Both devices are similar that they cover blemishes.

4.2 FGI on Skin AI, Beauty Tech, Skin Evaluation Model Setting

FGI result on primary analysis data through skin AI, beauty tech, and skin evaluation model case studies is as follows.

Services (devices) that prioritize on the analysis of consumer needs should be developed with the focus on the development of self-beauty device and product based on the technology of 4th Industrial Revolution Era. In order to develop products with practicality, convenience, as well as effectiveness, extensive skin analysis big data needs to be constructed.

Opinions on the necessity of basic learning DB through large grouping were accepted from the population of labeling and modeling for standardization of collected DB and objection handling. Also, some opinions suggest that, although it is widely used in various industries, some questions of Baumann Skin Type are focused on Caucasians that they do not reflect racial elements, and readjustment of questionnaire as well as continuous follow-up studies are needed. Through these FGI opinions, an additional analysis was attempted using studies and reports, which analyzed Korean skin using Baumann Skin Type.

Keywords were extracted by categorized skin diagnosis model by skin type and professional FGI was conducted. The result showed that reconstruction of skin diagnosis model evaluation index based on Baumann Skin Type is the most effective as it was used the most in skin researches and is the most generally used now. In addition, skin measurement in consideration of genetic and environmental factors for Korean skin would enable more appropriate skin analysis. However, this study was limited to case study and FGI, and additional studies on skin model for Asians are required through additional surveys, skin measurement experiments, etc.

4.3 Skin Types and Characteristics of Koreans Through Case Study

The result of analysis of skin condition survey by Foundation of Korea cosmetic Industry Institute on men and women living in Korea between 2011 and 2019 as a part of International Skin Characteristics Data Bank Construction Project is as follows [19]. Women's skin type appeared in the order of: combination, dry, normal, oily, and skin type survey through BST showed the highest ratio in DRPW type among 16 skin types (Table 5).

Table 5. The result of analysis of skin condition survey of female

Period	N	A primary classification of skin types	Baumann skin type	Type of skin concern	References
2011 7.19–8.12	150	Combination > Dry > Normal > Oily	Survey not in combination	Blemish > Pore > Skin Tone	[19]
2012 3.19 –5.12	200			Blemish > Wrinkles > Elasticity	
2013 1.14– 2.15	200			Blemish > Wrinkles > Skin Tone	
2014 3.3–3.14	106	Dry > Combination >Normal >Oily	DSPW > DSPT	Wrinkles > Dryness > Blemish	
2015 2.26–3.16	111	Combination > Dry > Normal > Oily	DRPT	Blemish > Elasticity	
2016 7.7 –7.22	122		DRPW > ORPT	Blemish > Skin Tone > Pore	
2017 6.26 – 7.7	107	Dry > Combination > Normal > Oily	DRPW > DRNW	Wrinkles > Blemish > Pore	
2018 8.13 –28	110		DRPW, DRNT > DRPT	Blemish > Wrinkles > Freckles	
2019 8.26– 9.6	106	Combination > Dry > Normal > Oily	DRPW > DRNT	Blemish > Elasticity > Pore	
Integration result		Combination > Dry > Normal > Oily	DRPW	Blemish > Wrinkles > Pore	

Age: 20–59.

Men's skin type appeared in the order of: normal, dry, oily, combination, and skin concerns included sebum, wrinkles, and pores. Skin type survey through BST showed the highest ratio in ORNT and DRNT types (Table 6).

Table 6. Dermatological and survey report of Korean male

Period	N	A primary classification of skin types	Baumann skin type	Type of skin concern	References
2011 7.19–8.12	150	Oily > Combination > Normal > Dry	Survey not in progress	Sebum > Pore > Acne	[19]
2012 3.19–5.12	200	Dry > Oily > Combination > Normal		Wrinkles > Sebum > Dryness	
2013 1.14–2.15	200	Normal > Dry = Oily > Combination		Wrinkles > Sebum > Blemish	
2014 3.3 – 3.14	110	Dry > Normal > Oily > Combination	OSNT > DRNT	Dryness > Wrinkles > Sebum	

(*continued*)

Table 6. (*continued*)

Period	N	A primary classification of skin types	Baumann skin type	Type of skin concern	References
2015 2.26–3.16	111	Dry > Norma > Combination > Oily	DSNW	Dryness > Wrinkles > Pore	
2016 7.7–7.22	114	Normal > Oily = Dry > Combination	ORNT > DRNT	Sebum > Wrinkles > Pore	
2017 6.26–7.7	107	Normal > Dry > Oily > Combination	ORNT > DRNT		
2018 8.13–28	110	Normal > Oily = Dry > Combination	ORNT > DRNT	Sebum > Wrinkles > Pore = Dryness	
2019 8.26– 9.6	108	Normal > Oily > Combination > Dry	ORNT > DRNT	Sebum > Pore > Dryness	
Integration result		Normal > Dry > Oily > Combination	ORNT > DRNT	Sebum > Wrinkles > Pore	

Age: 20–59.

The conformity between the result of Baumann Skin Type Questionnaire (BSTQ) on women who visited dermatologists and interviews with dermatologists was not high. Also, compared to other studies, the skin type tended to further categorize into non-pigmented (N) and sensitive (S) with the use of BSTQ, so accurate skin type needed to be determined through additional consultation with an expert [20]. In the study with 1000 Korean women, skin types appeared in the order of: OSNT, DSNT, DRNT, and OSNW (Table 7). Due to significant differences in age, region, drinking habit according to smoking, occupation, blood type, and UV exposure, this requires personalized care [21]. The dominant skin type in the study with 1000 Korean men was OSNW, and skin type distribution was distinct based on age and region, suggesting that skincare in consideration of environmental factors was necessary [22] .

Table 7. Analysis of Korean skin types using Baumann skin

Period	N	Period	Skin type	Reference
2014.8–2015.7	202	Visited the dermatologic clinic of our hospital (19–64 years of age)	Evaluation of Questionnaire: Dry > Oily > Sensitivity BST: DSNT, OSNT, DSNW	[20]
			Interview with a dermatologic specialist: Dry > Oily > etc. (Normal, Combination) > Sensitivity BST: DRPT, DRNT, OSNT	

(*continued*)

Table 7. (*continued*)

Period	N	Period	Skin type	Reference
n.d	1000	Healthy Korean Women Volunteers Without Skin Diseases	OSNT, DSNT, DRNT, OSNW	[21]
2018. 6–12	1000	Korean men divided equally by age and region (20 –60 years of age)	OSNW, DSNW, DRNT, ONST	[22]

Based on the analysis of Korean skin type using BST, women were DRPT type, appearing in the order of D > R = T > P, and men were DRNT type, appearing in the order of N > T > R > D.

5 Conclusion

As social values are facing a transition into a form of convergence industry with the innovation of the 4th Industrial Revolution, beauty service industry is showing joint growth of products and services. This study conducted case analyses of algorithm-based beauty tech products using skin information big data in order to examine the trend in beauty industry. Through this process, it aimed to suggest the direction of big data-based beauty tech, which examines the trend in beauty industry and satisfies consumers' needs. As the overall beauty industry is paying attention to the reinforcement of beauty tech service, as a combination of beauty and technology, strategies are needed to reinforce competitiveness of beauty industry. Since information delivery, channel, product, and beauty service beauty tech provides and suggests services and products based on big data, collection of diverse and extensive data based on skin characteristics would be more important than anything. This can improve the service accuracy and provide competitive beauty tech services and products. In addition, with regard to the skin model for skin analysis in big data-based beauty tech, a standardized skin evaluation model that can reflect Koreans' genetic factors, environmental factors, and cultural factors should be suggested in consideration of the expert group's opinion, which stated that some part of Baumann Skin Type Test is not appropriate for Koreans, and another opinion that an evaluation index for Koreans is needed.

Stable skin analysis system can be established through this and contribute to the rapid growth of K-beauty tech, which shows an average of 10% annual growth, as opposed to an average of 6% global annual growth. Therefore, continuous updates on skin information data needs to take place for more accurate suggestions. Also, consumers' needs and preferences need to be determined frequently to provide personal curation and customized services and cosmetics as a means to reinforce competitiveness. Customized products and services using AI will satisfy diversifying consumer needs and provide highly effective satisfaction. Continuous data update and research and development need to take place, so tailored and diversified beauty tech based on reliability and accuracy can bring advancement of beauty industry.

References

1. Park, M.S., Lee, D.H., Choi, J.A.: Analysis on the industrial linkages between manufacturing and service sector in Daegu and Gyengbuk Region. The Korean Region. Dev. Assoc. **29**(1). 99–120 (2017)
2. Son, J.O.: Personalized Recommendation and Search System Based on Beauty Big Data Considering User Preferences and Item Reliability (2019)
3. Lee, J., Bang, J.: A case study on the development of new brand concept through big data analysis for a cosmetics company. Knowl. Manag. Res **21**(3), 215–228 (2020)
4. http://jmagazine.joins.com/economist/view/329118
5. The 4th Industrial Revolution. https://ko.wikipedia.org/wiki/%EC%A0%9C4%EC%B0%A8_%EC%82%B0%EC%97%85_%ED%98%81%EB%AA%85#:text=%EC%A0%9C4%EC%B0%A8%20%EC%82%B0%EC%97%85%20%ED%98%81%EB%AA%85(%E7%AC%AC%E5%9B%9B%E6%AC%A1%20%E7%94%A3%E6%A5%AD%20%E9%9D%A9%E5%91%BD,%EB%A1%9C%20%EC%A4%91%EC%9A%94%ED%95%9C%20%EC%82%B0%EC%97%85%20%EC%8B%9C%EB%8C%80%EC%9D%B4%EB%8B%A4
6. Kim, G.M.: A study on the changes of the 4th industrial revolution and global beauty market. J. Cult. Prod. Des. **50**, 221–231 (2017)
7. https://www.mk.co.kr/news/it/view/2020/08/855469/
8. Biotimes.: http://www.biotimes.co.kr
9. Baumann, L.: The Skin Type Solution, Bantam Books, New York (2007)
10. Yoon, Y.S., Kim, M.S.: New Skin Management. Cover Book Publishing, Seoul (1996)
11. Kim, M.S.: Skin Care. Preface, pp. 170–173 (2001)
12. https://www.laroche-posay.us/my-skin-track-uv-3606000530485.html
13. https://porescan.com/#/
14. http://www.lulu-lab.com/
15. https://www.neutrogena.com/skin360app.html
16. www.econovill.com/news/articleView.html?idxno=354344
17. https://news.mt.co.kr/mtview.php?no=2019123009063718917
18. https://www.hiic.re.kr/vol-1-ces-2020-%EB%B7%B0%ED%8B%B0%ED%85%8C%ED%81%AC-%EC%9D%B8%EA%B3%B5%EC%A7%80%EB%8A%A5-%EA%B0%9C%EC%9D%B8%ED%99%94/
19. Foundation of Korea cosmetic Industry Institute.: The result of analysis of skin condition survey (2011–2019)
20. Choi, J.Y., Choi, Y.J., Nam, J.H., Jung, H.J., Lee, G.Y., Kim, W.S.: Identifying skin type using the Baumann skin type questionnaire in Korean women who visited a dermatologic clinic. Korean J. Dermatol. **54**(6), 422–437 (2016)
21. Ahn, S.K., et al.: Baumann skin type in the Korean female population. Ann. Dermatol. **29**(5), 586–596 (2017)
22. Lee, Y.B., et al.: Baumann skin type in the Korean Male population. Ann. Dermatol. **31**(6), 621–630 (2019)

Technical Countermeasures Against Drone Communication Vulnerabilities

Wonhyung Park[1] and Hoo-Ki Lee[2](✉)

[1] Department of Information Security Engineering, Sangmyung University, Cheonan, Korea
whpark@smu.ac.kr

[2] Department of Cyber Security Engineering, Konyang University, Nonsan, Korea
hk0038@konyang.ac.kr

Abstract. With the Fourth Industrial Revolution, the use of drones is rapidly expanding into various fields; drones are now available to a variety of consumers in the agricultural, industrial, security, firefighting, and private sectors. However, along with the proliferation of various types of drones, there has been an upsurge in hacking resulting from a lack of security guidelines governing drone communication. In particular, the malicious hacking of airborne drones can escalate through extortions or crashes. Although many attempts at hacking drones have been scuttled, there are no clear guidelines on how to respond in such situations. In this study, we examine countermeasures against drone communication (Wi-Fi, Bluetooth, and GPS) vulnerabilities.

Keywords: Communications vulnerabilities · Drone · Fourth Industrial revolution · Industrial security · Information security

1 Introduction

Drones began to appear in various airspaces in the early 2000s. The word "drone" was originally used to describe the humming of bees, and in the early days, drones were developed as unmanned aerial military vehicles [1]. As drones came into wider use, they began to be utilized in sectors such as agriculture, manufacturing, security, and firefighting. With the expansion into private use, the number of scenarios that could endanger society is steadily increasing in the form of privacy violations and the threat of terrorism owing to the unchecked use of drones. Drones count among the main technologies of the Fourth Industrial Revolution and are likely to be continuously and significantly developed [4]. Most drones communicate at the 2.4 GHz [Wi-Fi, Bluetooth] and 5.8 GHz [Wi-Fi] frequency bands [5]. When the signal transmitted to maintain the drone flight—such as GPS—drone-controller communication—is compromised or cut off, drones become exposed to attacks.

Cases of terrorism involving drones have been reported in various countries; in 2011 for instance, the U.S. military's RQ-170 Sentinel was reportedly hacked by Iran. In another example of aerial warfare, a Saudi oil complex was attacked by several drones in 2019. In light of such incidents, we examine the vulnerability of drones released to the market with respect to interference of wireless communication and suggest relevant countermeasures.

R. Lee (Ed.): ICIS 2021, SCI 1003, pp. 119–129, 2022.
https://doi.org/10.1007/978-3-030-90528-6_11

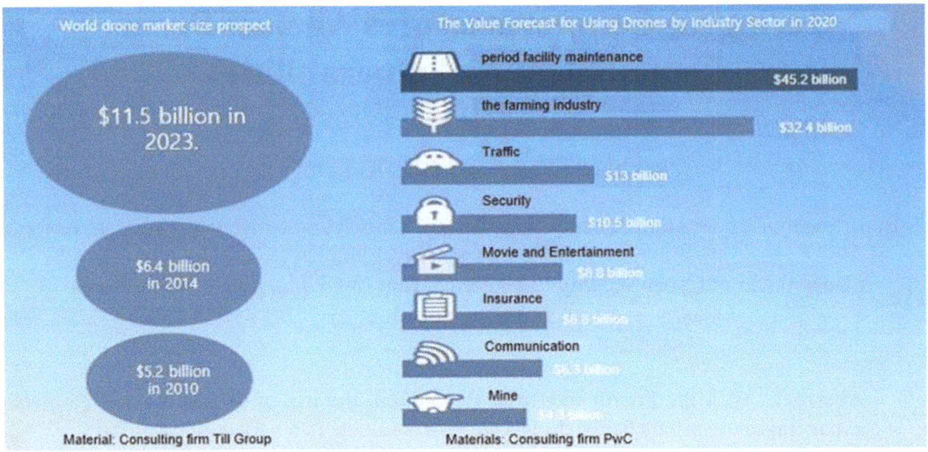

Fig. 1. The size of the drone market

2 Related Work

Drones typically rely on 2.4 GHz and 5.8 GHz wireless frequency bands for signal transmission. Accordingly, modules related to remote drone control—such as transmitting images per the pilot's request, geographical positioning, and automatic posture control—have been developed [6]. Drones operate in Wi-Fi, Bluetooth, and GPS frequency bands as shown in Table 1 below.

Table 1. Drone communication frequency band-

	Frequency
Wi-Fi	2.4 GHz, 5.8 GHz
Bluetooth	2.4 GHz
GPS [7]	L1 (1575.42 MHz)
	L2 (1227.60 MHz)

Jamming, replay, and spoofing attacks can be executed in the corresponding frequency band.

2.1 Jamming Attack

A jamming attack is a form of network attack that severs the drone's wireless connection by transmitting a stronger signal compared to the Wi-Fi or Bluetooth frequency band between the drone and the controller, hence paralyzing communication [8]. Similarly, by sending a signal stronger than the GPS signal received by the drone, jamming can prevent the drone from coordinating geographical information [9].

2.2 Spoofing Attack

The dictionary defines "spoofing" as hoaxing or tricking. A spoofing attack converts a normal signal into an abnormal signal. Drones use GPS to get geographical flight information. By sending a forged GPS signal in the direction of the drone, a spoofing attack can change the current geographical information of the drone, effectively hijacking it [10].

2.3 Replay Attack

A replay attack is also known as a retransmission attack. In this form of network attack, the attacker takes control of a communication signal between the controller and the target (drone); the attacker then retransmits the communication signal to the target of attack. Communication signals between the drone and its controller typically contain information for taking off and landing as well as geographical and positioning information. Upon accessing such information and screening the valid signals, the attacker can remotely control the drone by retransmitting these signals to the drone [11].

3 Drone Communication Vulnerability Anaysis

This section describes a drone-vulnerability attack scenario.

3.1 Drone Communication Vulnerability Attack Scenario

This scenario simulates the forced landing of a drone consequent hijacking by exploiting loopholes in the drone's communication network. The attack scenario is as follows:

1) Selection of targeted drones.
2) Inspection of targeted drones (communications, modules, etc.).
3) Comparison of previous and current attacks.
4) An investigation into the attack using HackRF One and analysis from another perspective other than radiofrequency (RF) communications.
5) Initiating the drone vulnerability attack
6) Forced landing and control deauthorization.

The specification of the drone used for simulation and its corresponding configuration of attack environment are as follows:

- Attack Target Drone
- Manufacturer: DJI
- Product name: Phantom 4 Pro, Spark
- File system: FAT32
- Communication: Wi-Fi (2.400 GHz–2.483 GHz, 5.725 GHz–5.825 GHz), GPS

- Configure Attack Environment
- H/W: PC, HackRF One
- System: Windows 10, Linux (Kali, Ubuntu)
- Usage Program: GNU-Radio (RF Communication Frequency Analysis and Output), SDR# (RF Communication Frequency Analysis), Advanced Port Scanner (Network Port Scan), JD-GUI (Java Program Decompilation) etc.

3.2 Basic Drone Communication Analysis

When the communication between the controller and the drone is analyzed, the controller can confirm when the malicious signal appeared (ON) and disappeared (OFF), as shown in Fig. 2. In the case of an ON signal, the 2.4 GHz band was set to 2.401–2.402 GHz, and the 5.8 GHz band was set to 5.726–5.727 GHz [12]. OFF signals randomly confirmed that these appeared only in the 2.4 GHz and the 5 GHz band

Fig. 2. Signal generation in controller: ON/OFF

When the drone was turned on, it was confirmed that each channel used for drone–controller communication was consistently transmitting 2.4 GHz and 5 GHz communication signals, as in Fig. 3 and Fig. 4.

Fig. 3. Drone on signal generated

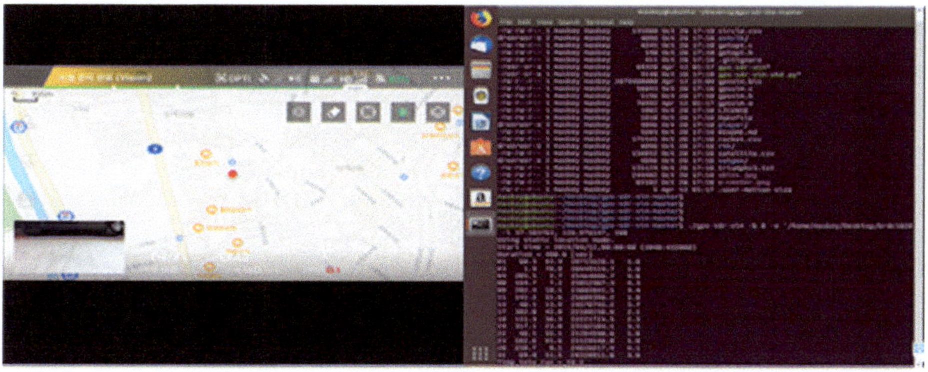

Fig. 4. Location of drones before GPS spoofing attacks

3.3 Drone Communication Vulnerability Verification

3.3.1 GPS Spoofing Attack

A GPS spoofing attack utilizes the drone's GPS module. The drone has a setting for a restricted flight zone, and it was confirmed that the drone could not take off within the restricted area; moreover, the drone will automatically land if airborne in restricted airspace. During the attempt at a GPS spoofing attack, the logs on the drone confirmed an automatic landing after a countdown, as shown in Fig. 5.

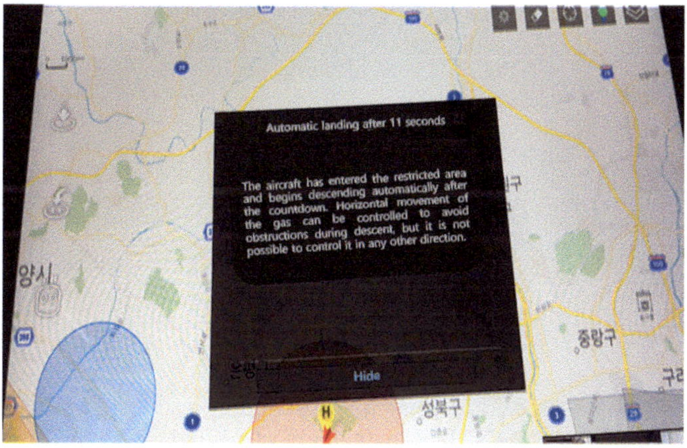

Fig. 5. Automatic landing of drones after GPS spoofing attacks

3.3.2 Account Exposure for FTP Communication

A data packet was captured and verified as shown in Fig. 6 and Fig. 7 by connecting to the drone network.

A corresponding network was using the network as shown in Table 2.

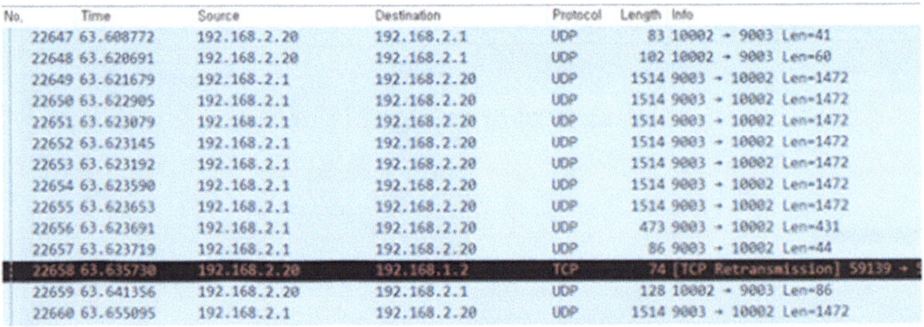

Fig. 6. Drone controller network packet

Fig. 7. Drone network port scan

Table 2. Drone Wi-Fi network environment

Sortation	IP	Port
Drone	192.168.2.1	9003
Controller	192.168.2.20	10002

A port scanner was used to identify the ports being used in drones (Fig. 8). The BusyBox ftpd D-Link DCS-932L IP-Cam camera identified an open FTP port. The port is possibly a port for video transmission between the user and the drone.

As a result of decompiling the "DJI" application and analyzing it, we confirmed that DJI performs various functions using the FTP protocol (Fig. 8) and the Package dji.pilot.publics.control.upgrade. Using the "filot.publicics.control.upgrade," the account used to log in with the code related to upload was identified (ID: anonymous; PW: None). Also, as shown in Fig. 9 we confirmed that a proxy using FTP was used as a package (p661it.sauronsoftware.ftp4j.connectors, and ID: anaonymous; PW: ftp4).

```java
public void mo43244d() {
    String ftpIP = this.f255631.getFtpIP();
    int ftpPort = this.f255631.getFtpPort();
    String dir = this.f255631.getDir();
    C6650a.m27768a(this.f25553b, "upload sendWithFtp ip=" + ftpIP + " port=" + ftpPort);
    C6650a.m27768a(this.f25553b, "upload sendWithFtp dir=" + dir);
    String[] split = dir.split("/");
    String str = split[split.length - 1];
    String replace = dir.replace(str, "");
    File file = new File(this.f25575x);
    File file2 = new File(file.getParent() + "/" + str);
    file.renameTo(file2);
    final FTPClient fTPClient = new FTPClient();
    try {
        fTPClient.connect(ftpIP, ftpPort);
        fTPClient.login("anonymous", "");
        if ("".equals(replace)) {
            replace = "/";
        }
        fTPClient.changeDirectory(replace);
        fTPClient.getConnector().setConnectionTimeout(20000);
        fTPClient.upload(file2, (FTPDataTransferListener) new FTPDataTransferListener() {
            public void started() {
                C6650a.m27768a(C6706d.this.f25553b, "upload sendWithFtp started");
            }

            public void transferred(int i) {
                long unused = C6706d.this.f25560i = C6706d.this.f25560i + ((long) i);
                if (C6706d.this.f25560i % StatService.BYTES_IN_MEGA == 0 || C6706d.this.f25560i >= C6706d.this.f25559h) {
                    C6706d.this.f25566o.mo43293t();
                }
            }

            public void completed() {
                C6650a.m27768a(C6706d.this.f25553b, "upload sendWithFtp completed");
                C6706d.this.m28188m();
                C6706d.this.m28194p();
                try {
```

Fig. 8. Package: dji.pilot.publics.control.upgrade My account exposed

```java
public FTPProxyConnector(String str, int i, String str2, String str3) {
    super(true);
    this.style = STYLE_SITE_COMMAND;
    this.proxyHost = str;
    this.proxyPort = i;
    this.proxyUser = str2;
    this.proxyPass = str3;
}

public FTPProxyConnector(String str, int i) {
    this(str, i, "anonymous", "ftp4j");
}
```

Fig. 9. Package: p661it.sauronsoftware.ftp4j.connector My account exposed

3.3.3 Relay Attack

Upon connecting the drone to the controller, we confirmed that a signal was being transmitted, as shown in Fig. 10. We then attempted a replay attack by capturing the signal; however, we verified that the signal was safe from attack because there were external security measures in place.

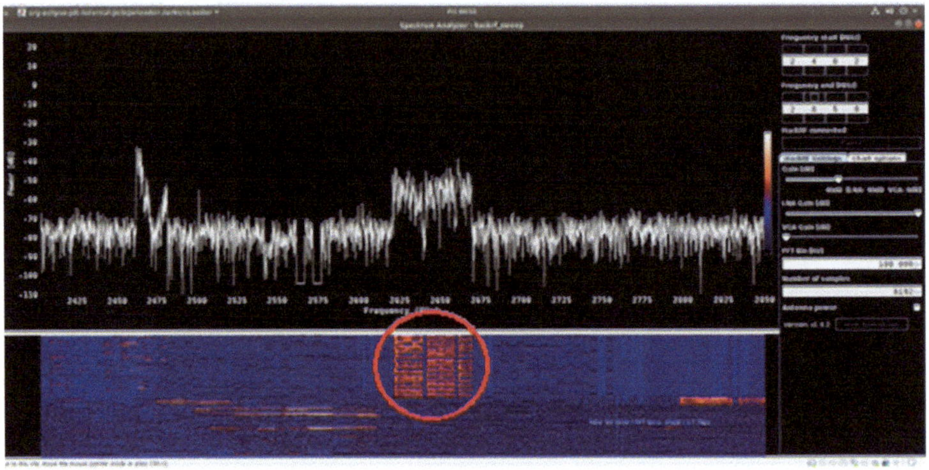

Fig. 10. Controller drone connection

3.3.4 Expected Drone Vulerabilities

In Fig. 11 BusyBox—of BusyBox ftpd D-Link DCS-932L IP-Cam Camera—is a program that collects frequently used commands in UNIX systems. It provides an environment suitable for embedded systems that require only minimal resources [13]. The use of BusyBox was confirmed to be used in drones as an alternative to DJI [14]. Consequently, FTP vulnerabilities associated with BusyBox were found in many areas, including CVE-2017-3209 and OWASP announcements. The drone is also susceptible to attack courtesy of the inadequacies of BusyBox [15].

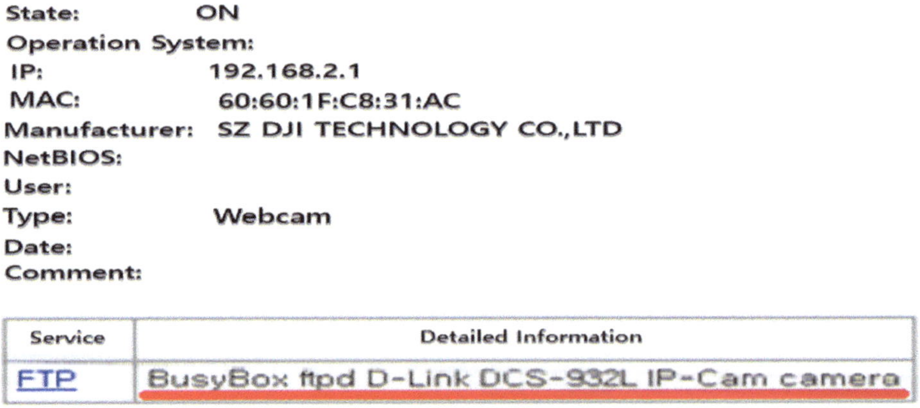

State: ON
Operation System:
IP: 192.168.2.1
MAC: 60:60:1F:C8:31:AC
Manufacturer: SZ DJI TECHNOLOGY CO.,LTD
NetBIOS:
User:
Type: Webcam
Date:
Comment:

Service	Detailed Information
FTP	BusyBox ftpd D-Link DCS-932L IP-Cam camera

Fig. 11. Information displayed during port scanning

4 Drone Communication Vulnerability Response Plan

As drones continually diversify, a minimum/standard level of security is required for standard off-the-shelf drones; essentially, only expensive commercial models may require extensive countermeasures. Most drones have controls and video acquisition functions utilizing Wi-Fi or Bluetooth communication [16]. There are also drones that have functions such as GPS. Communication networks used by drones are now open to attack. Therefore, drones across the board require the application of a minimum-security threshold. This paper proposes a solution to the vulnerability.

4.1 GPS Spoofing Attack

As a countermeasure against GPS spoofing attacks, two conditions must be met; the return-to-home (RTH) function and an electronic compass should be built in. The RTH function stores GPS coordinates when the drone takes off and returns to the take-off position autonomously. Using this function, the RTH coordinates and the straight-line distance of the active GPS coordinates are calculated as shown in Fig. 12 and 13. when the drone is out of the communication range, this is considered a GPS spoofing attack.

Drones use an electronic compass to store RTH flight coordinates, converting these to electronic compass-based coordinates as opposed to GPS coordinates when a GPS attack is detected, and then reverting to RTH values.

$$R(Earthradius) = 6378.135km$$

$$\cos(Degrees) = \frac{RTHCoordinatelatitude^{\circ} + Currentcoordinatelatitude^{\circ}}{2}$$

$$C = \cos(Degrees)^{*}\left(\frac{2\pi R}{360}\frac{2\pi R}{360}\right)$$

$$D = \frac{2\pi R}{360}$$

$$Distance = \sqrt{((Degrees^{*}C) + (Min^{*}(\frac{C}{60}) + (Sec^{*}(\frac{(\frac{C}{60})}{60})))^{2} + ((Degrees^{*}D) + (Min^{*}(\frac{D}{60}) + (Sec^{*}(\frac{(\frac{D}{60})}{60})))^{2}}$$

Fig. 12. When GPS coordinates are in the degrees, minutes, and seconds format

```
private static double distance(double RTH_Lat1, double RTH_Lon1, double
DJI_Lat2, double DJI_Lon2){
        double longitude = RTH_Lon1 - DJI_Lon2;
        double distance_lat_lon = Math.sin(Math.toRadians(RTH_Lat1)) *
Math.sin(Math.toRadians(DJI_Lat2)) + Math.cos(Math.toRadians(RTH_Lat1)) *
Math.cos(Math.toRadians(DJI_Lat2)) * Math.cos(Math.toRadians(longitude));

        distance_lat_lon = Math.acos(distance_lat_lon);
        distance_lat_lon = Math.toDegrees(distance_lat_lon);
        distance_lat_lon = distance_lat_lon * 60 * 1.1515;

        distance_lat_lon = distance_lat_lon * 1609.344;
        return (distance_lat_lon);
}
```

Fig. 13. Java code with decimal GPS coordinates

4.2 Account Exposure for FTP Communication

An android application can easily check the source code of the application with the program JD-GUI as shown in Fig. 14.

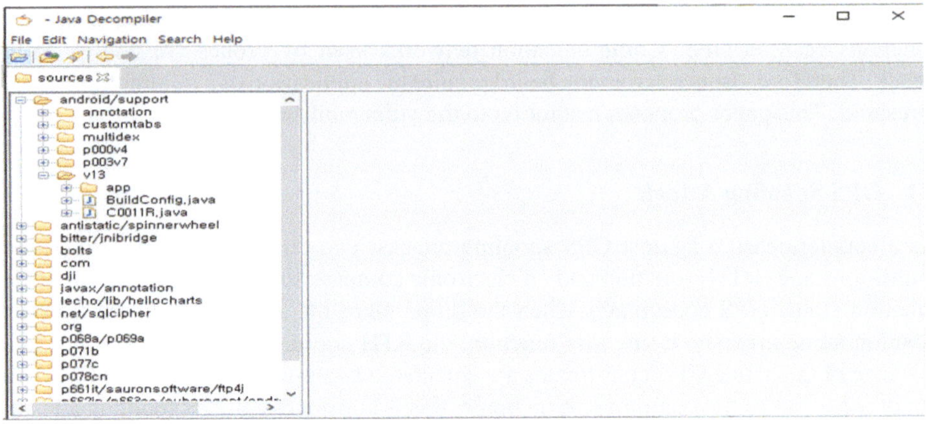

Fig. 14. Java Decompiler jd-gui

Important information (ID, PW, etc.) related to drone communication is encrypted or obfuscated to prevent the exposure of vital information when analyzing the application. Similarly, when the drone and the controller are connected, a dynamic connection method is required to transmit appropriate GPS-based random text using the source code Fig. 15 and subsequently generate random PW with the corresponding text value.

```
public class FTP_pw_enc {
    public static void main(String[] args) throws NoSuchAlgorithmException
                , NoSuchPaddingException, InvalidKeyException, IllegalBlockSizeException
                , BadPaddingException, UnsupportedEncodingException {


        String encryptionKey = "[16byte 키 값]";
        String target = RTH_KEY(RTH_Lat1, RTH_Lon1);
        Cipher cipher = Cipher.getInstance("AES");


        SecretKeySpec                    secretKeySpec                    =                    new
SecretKeySpec(encryptionKey.getBytes(),"AES");
        cipher.init(Cipher.ENCRYPT_MODE, secretKeySpec);
        byte[] encryptBytes = cipher.doFinal(target.getBytes("Base64"));
    }
    private static String RTH_KEY(double RTH_Lat1, double RTH_Lon1){
        double key1 = RTH_Lat1 - RTH_Lon1;
        if (key1 < 0){
                key1 = Math.abs(key1);
        }
        double temp_key;
        String in_temp;
        temp_key = (key1 - (int)key1)*1000000;
        in_temp = Integer.toString((int)temp_key);

        return (in_temp);
    }
}
```

Fig. 15. Create Drone-side Random PW

5 Conclusion

In this paper, we duly demonstrated the vulnerability to ground drones by exploiting flaws in wireless communication networks. We also explored applications that may be used as countermeasures against covert hacking of drones. Through application analysis, information such as networks and accounts used by the drone were obtained, and while vulnerabilities in drones using BusyBox were not identified, we hold that attacks on drone systems thrive on unregulated security. We also confirmed that drones using geographical positioning are exposed to GPS jamming and spoofing. As drone use burgeons in the Fourth Industrial Revolution, it is imperative to study drone vulnerabilities and attack methods with a view to developing appropriate security countermeasure.

References

1. Naver Knowledge Encyclopedia, Drone (2018)
2. Young-wook, L.: The direction of development of drones in Korea during the 4th Industrial Revolution. Converg. Secur. J. **18**(5), 3–10 (2018)
3. In-soo, J.: Analysis of vulnerabilities in drones using Wi-Fi. Comput. Inf. Soc. Korea **25**(1), 219–222 (2017)
4. Seong-hwa, S.: Overview and issues of drone radio communication. J. Korea Telecom. Assoc. **33**, 93–99 (2016)
5. Joo-hwan, S.: Deception and response using the drone's wireless network security vulnerability. Information and Communication Society of Korea. **5**, 327–330 (2017)
6. Myung-soo, K.: Trend of vulnerability analysis and response technology of unmanned vehicle drones. J. Soc. Inf. Prot. **30**(2), 49–57 (2020)
7. Bok-seop, S.: A study on the performance improvement of PRC regeneration algorithm in multi-DGPS dynamic environment. Master's Degree thesis, Hanbat University School (2007)
8. Sârbu, A.: Wi, FI jamming using software defined radio. In: Proceedings of the International Conference on The Knowledge-Based Organization, pp. 162–166 (2020)
9. Nurkic, L.: Difficulties in achieving security in mobile communications. In: Encarnação, J.L., Rabaey, J.M. (eds.) Mobile communications. IFIP — The International Federation for Information Processing, pp. 277–284. Springer, Boston, MA (1996). https://doi.org/10.1007/978-0-387-34980-0_28
10. Hwa, J.I.: GPS Jamming Technique and Anti-Jamming GPS Technology. Korea Assoc. Inf. Commun pp. 573–575 (2015)
11. Joon-woo, K.: Anti-drone using GPS-spoofing. Korean Soc. Inf. Process. **27**(5), 338–341 (2020)
12. Seung-yeon, C.: Analyze replay attack vulnerabilities in RF communications environments. Master's thesis, Korea University's Graduate School of Information Security (2017)
13. Young-hyuk, Y.: Designing and implementing high-efficiency slot antenna for metal laptop dual-band WiFi MIMO. J. Electr. Soc. **67**(10), 1138–1343 (2018)
14. BusyBox. https://www.busybox.net/about.html.
15. Ga-yeon, R.: A drone forensics study using linux-based BusyBox. Korean Soc. Inf. Process. **24**(1), 273–275 (2017)
16. Astaburuaga, I.: Vulnerability analysis of AR. drone 2.0, an embedded Linux system. In: IEEE 9th Annual Computing and Communication Workshop and Conference (CCWC), vol. 1, pp. 666–672 (2019)
17. Ho, S.C.: A study on the security vulnerability analysis and countermeasures of drones. Inf. Commun. Soc. Korea **5**, 355–358 (2016)

Localization in LoRa Networks Based on Time Difference of Arrival

Ioannis Daramouskas, Dimitrios Mitroulias, Isidoros Perikos(✉), Michael Paraskevas, and Vaggelis Kapoulas

Computer Technology Institute and Press "Diophantus", 08544, 26504 Patras, Greece
perikos@ceid.upatras.gr

Abstract. In this paper, we examine the localization capabilities of LoRa networks in terms of the localization error. A LoRa network in the area of our University was created with precisely synchronized clocks and TDoA capabilities. The localization performance of different methods is thoroughly examined under grouping or not of the messages that were sent by the sensor. In the first case, we utilize sets of k messages to calculate the average of TDoAs of these messages and use the average value for every pair of gateways to perform the localization. In the other case, localization is performed by using each one signal sent by the target sensor, and the efficiency and the performance of the algorithms are assessed. Various experiments were performed to assess the localization capabilities of LoRa technology in the real-world setup. The results of the experiments are quite interesting. First, the results indicate that the localization error is greatly affected by the noise of the recorded timestamps of the base stations. The noise results in underestimating and overestimating the pseudo-ranging something that greatly affects the performance of the algorithms. Also, the results point out that the grouping of the received messages is quite useful in minimizing the localization error.

Keywords: Localization · LoRa · TDoA · Particle Swarm optimization · Least squares

1 Introduction

LoRa technology is a recent technology belonging to the Low Power and Wide Area networks offering a wide range of unique features [16]. It consists of two parts, LoRa (physical level) and LoRaWAN [6]. LoRa is a configuration technique for a specific wireless spectrum, while LoRaWAN is an open protocol that allows IoT devices to use LoRa for communication [5]. It is based on the cloud, designed and supported by the LoRa Alliance, which allows devices to communicate wirelessly with LoRa. In essence, LoRaWAN takes LoRa wireless technology and adds a networking element to it, while at the same time integrating node authentication and data encryption for security [4]. Basic foundations of the LoRa approach cuts ages through messaging up to 10 km, in Line of Sight (LOS) situations. LoRa technology uses 868/915 MHz which can be used internationally.

R. Lee (Ed.): ICIS 2021, SCI 1003, pp. 130–143, 2022.
https://doi.org/10.1007/978-3-030-90528-6_12

Due to the low power consumption and the long-range communication, Low Power Wide Area Networks (LPWAN) standards like LoRaWAN are implemented in IoT devices to enable efficient and quite low-power wireless communication [9]. Moreover, the characteristics of this wireless communication link can also be used for localization purposes [10]. The analysis of the received signal strength or the Time of Arrival (ToA) at the receiving gateways, can include approaches like multilateration, triangulation, or pattern matching to specify the location of the transmitter [8].

This work is conducted under a research project named [Removed due to blind review procedure]. The project aims to design and develop a system for identifying and rescuing individuals, especially those belonging to population groups with a particularly high probability of being lost. The central part of the system is a modular "wearable (portable)" device, while the positioning and statusing algorithms of the person carrying the device will play an important role in its effectiveness. The wearable device will offer the possibility of basic communication with base stations, that can be located many kilometers away from the device's location. The wearable will use low energy consumption communication and long reach protocols (LoRa). The rhythm or the force of the broadcasted data packets will adjust to the special conditions of application, so that battery life can last for days or even weeks. This wearable system is considered to be best suited for use in the following scenarios. In Fig. 1 we can see an example of a Lora network in a city where the base stations are placed at the top of the buildings and the wearable device the woman possess can communicate with each the base stations and each base station transmits the data to a cloud service.

Fig. 1. An example LoRa network.

In this paper, we examine the localization capabilities of LoRa networks in terms of the localization error. A LoRa network in the area of our University was created which covers an area of 8 km × 6 km and consisted of 4 base stations with precisely synchronized clocks and Time Difference of Arrival (TDoA) capabilities. The localization performance of different methods is thoroughly examined under grouping or not of the messages that were sent by the sensor. In the first case, we utilize sets of k messages to calculate the average of TDoAs of these messages and use the average value for every pair of gateways to perform the localization. In the other case, localization is performed by using each one signal sent by the target sensor, and the efficiency and the performance of the algorithms are assessed. Various experiments were performed to assess the localization capabilities of LoRa technology in the real-world setup. The results of the experiments are quite interesting. First, the results indicate that the localization error is greatly affected by the noise of the recorded timestamps of the base stations. The noise results in underestimating and overestimating the pseudo-ranging (the difference between the time the signal arrives first to a base station minus the time the signal arrives at the others multiplied with the propagation speed) something that greatly affects the performance of the algorithm. Also, the results point out that the grouping of the received messages is quite useful in minimizing the localization error.

The remainder of the article is structured as follows: Sect. 2 surveys the literature and presents related works on LoRa localization with TDoA. Section 3 presents our work, the LoRa network created in the area of our University, and the methodology on localization with TDoA. Section 4 presents the experimental study while Sect. 5 presents the results collected and the main findings of the study. Finally, Sect. 5 concludes the article and provides directions that future work will examine.

2 Related Work

Localization constitutes a fundamental procedure with various applications on a wide spectrum of domains. Efficient localization is highly desirable, and many works study the efficiency of different methods and techniques in various contexts [2, 7].

Localization can be performed by utilizing the Received Signal Strength Indicator (RSSI) parameters [17]. In such a case, the location of a node in a network can be estimated with the RSSI metrics that are received by the corresponding gateways when a specific node conducts a signal transmission. Another method is based on the measurements of the Angle of Arrival (AoA). In such a context, the position of a node is estimated by a couple of base stations that are equipped with proper arrays of antennas [15]. The Angle of Arrival approach is suitable to be used in line-of-sight conditions [11, 12]. Time of Arrival (ToA) constitutes an approach that bases the localization on the actual time that a message was transmitted by the sensors and the actual time that it was received by each of the base stations. The ToA approach is greatly limited by the need for accurate synchronization between the base stations and the specific transmitting device. Time Difference of Arrival (TDoA) constitutes a quite potent localization approach [18]. It is based on the calculation of the difference between the time of arrival of a signal emanating from a transmitter to remote base stations. In such a context, the position of a node is estimated using 3 or more base stations that possess accurate time references.

Many methods utilize the TDoA metrics to perform localization in LoRa Networks. The work presented in [13] utilizes TDoA in an outdoor LoRa setup of 10×10 km and the authors study localization approaches like the Maximum Likelihood Estimation and the Multilateral Dissection. The authors use up to 19 base stations and report an estimation error of around 500 m. In the work presented in [14] the authors experiment with the TDoA localization in an outdoor context of 1.74×1.74 km. In total, four base stations are used, and authors report an average error in the localization procedure that is 100 m on average.

3 Methodology

In this section, we illustrate the work methodology and the methods that were used in our research. The implementation of the LoRa network is presented describing the gateways and their exact positions as well as the sensors used.

The LoRa network was implemented in our city in the area of our University Campus. Four MultiTechConduit IP67 LoRa base stations were placed and created an area of 8 km × 6 km (Fig. 2).

Fig. 2. The LoRa network we created in our area

The LoRa gateways that were used are the MultiTechConduit IP67 Base Station models that are created by the MultiTech. They are a ruggedized IoT gateway solution, specifically designed for outdoor LoRa public or private network deployments. The model is based on the Semtech v2.1 reference design, which uses the LoRaWAN protocol to perform Time Difference of Arrival (TDoA) calculations to deliver end-node location information in conjunction with a v2.1 LoRa Network Server. In the context of our LoRa Network, we used the MultiTech Conduit LoRa Gateway 4G Outdoors IP67 & Linux Geolocation Base Station models (Semtech based 2.1) with Ref MTCDTIP-L4E1-270L-868 as illustrated in Fig. 3.

Fig. 3. The MultiTech Conduit LoRa Gateways was used in the study.

The MultiTechConduit IP67 Base Station is a widely used model for localization in LoRa networks using TDoA measurements since it is capable to provide nanosecond timestamps. In addition, the Elsys ERS sensor is also used for the localization procedure. ERS sensor is illustrated in Fig. 3 and is a smart and professional LoRaWAN sensor for indoor climate measurements and can be used outdoors just for transmitting empty messages. It has a clean and modern design which makes it discrete in both business and home environments. ERS can measure temperature, humidity, light, and room activity. The sensor is highly configurable, and the user can change all parameters as they like. ERS works great as a sensor for smart building, dynamic workspaces, or climate control.

Fig. 4. ELSYS ERS sensor was used in the study.

The ERS is a LoRaWAN Certified sensor and is a quite suitable sensor to be used in LoRa networks. The sensor has been used for sending messages and collecting the corresponding timestamps from the gateways to facilitate the localization procedure based on the TDoA measurements.

3.1 Localization Based on TDoA

Time difference of arrival (TDoA) based on a group of sensor nodes with known locations has been widely used to locate targets. Assume that there are N ($N \geq 3$) gateways which also can be called base stations (BSs), to determine the position of the target. The coordinates of the base stations are known, $B_i = (x_i, y_i)$ in Cartesian coordinates. The target's sensor coordinates are $T = (x, y)$. Take the base station with the smaller time of arrival as a reference and assume that the times that the signal arrives t_i with propagation speed c (speed of light). The range difference between the reference base station and the other base stations can be mentioned as $r_{i,1}$. The range difference error is a zero-mean Gaussian with a known standard deviation.

$$r_{i,1} = c * (t_i - t_1) \tag{1}$$

In addition, $r_{i,1}$ is also calculated as:

$$r_{i,1} = d_{i,1} \pm n_{i,1} \tag{2}$$

so from Eqs. (1) and (2), we have that:

$$c * (t_i - t_1) = d_{i,1} \pm n_{i,1} \tag{3}$$

where

$$d_{i,1} = d_i - d_1 \tag{4}$$

In Eq. (4), d_i represents the distance between the target sensor and the base station i, and d_1 represents the distance between the target sensor and base station 1, which is the reference base station. The distances can be expressed as:

$$d_1 = \sqrt{(x - x_1)^2 + (y - y_1)^2} \tag{5}$$

$$d_i = \sqrt{(x - x_i)^2 + (y - y_i)^2}, \text{ where } i, \in [2, N] \tag{6}$$

So, the problem of obtaining results based on TDoA measurements is the problem of solving $N - 1$ equations of type (3). When dealing with TDoA measurements, at least 3 base stations are needed to estimate the location of the target sensor.

Localization Methods

To perform the localization based on TDoA in the context of our LoRa network, we implemented and used the Iterative nonlinear least-squares and the social learning particle swarm optimization which we present below.

Social Learning Particle Swarm Optimization

In the Social Learning Particle Swarm Optimization (SL-PSO) algorithm, the learning process of the particle is updated through the personal best position and global best position only. The process of learning and imitating the behavior of better individuals in a population is known as social learning, which can be widely discovered in social animals [1]. The SL-PSO algorithm [1] is executed on a sorted swarm, in which a particle can perform social learning. Social learning helps the particles to learn from any better particles, where a particle in the current swarm can learn and imitate the behavior of any better particles, known as demonstrators. Imitators are the particles that learn or imitate the behaviors of the demonstrators in the current swarm.

The cost function is the average of the squared differences of the real pseudo-ranging distances with the calculated particles' pseudo-ranging distances.

Iterative non-Linear Least Squares

In this section, 2-D target localization based on TDoA measurement is presented. Assume that there are N ($N \geq 3$) Base Stations (BSs), to determine the position of the target. The coordinates of the base stations are known, $B_i = (x_i, y_i)$ in cartesian coordinates. The target's coordinates are $T = (x, y)$.

Solve a nonlinear least-squares problem with bounds on the variables. Given the residuals f(x) (an m-D real function of n real variables) and the loss function rho(s) (a scalar function), least-squares finds a local minimum of the cost function F(x).

$$h = \begin{bmatrix} R_{2,1} - (R_2 - R_1) \\ \vdots \\ R_{N,1} - (R_N - R_1) \end{bmatrix} \tag{7}$$

R_i represents the distance between the *ith.* base-station and the solution in each iteration. $R_{i,1}$ represents the real data measurements as illustrated in Eq. 3.

$$G = \begin{bmatrix} \frac{(X_1-x)}{R_1} - \frac{(X_2-x)}{R_2} & \frac{(Y_1-y)}{R_1} - \frac{(Y_2-y)}{R_2} \\ \vdots & \vdots \\ \frac{(X_1-x)}{R_1} - \frac{(X_N-x)}{R_N} & \frac{(Y_1-y)}{R_1} - \frac{(Y_N-y)}{R_N} \end{bmatrix} \quad (8)$$

$$\Delta = \begin{bmatrix} \Delta_x \\ \Delta_y \end{bmatrix} = \left(G^T * Q^{-1} * G\right)^{-1} * G^T * Q^{-1} * h \quad (9)$$

The covariance matrix is denoted as Q, with each iteration Δx adds to x_0 and Δ_y to y_0. The whole process is repeated until the termination conditions are satisfied like a minimum distance between Δ_x and Δ_y or exceeding the number of iterations.

4 Experimental Study

In this section, we present the experimental studies conducted and the results collected. Our experiments are based on real data collected from a LoRa sensor in a LoRa network placed in the area near our University campus in four different locations. The real data gathered from the ELSYS Sensor will be presented. Three target positions were used where the target sensor was in a stationary position and was transmitting messages using a spreading factor set to 12, which allows transmission in longer distances but affects the energy of the signal. In Table 1, the distances between the base stations are presented while in Table 2 the distances between the target points and base stations are illustrated.

Table 1. Distances between the base stations in meters

	Base station 0	Base station 1	Base station 2	Base station 3
Base station 0	–	1018	5984	3078
Base station 1	1018	–	7003	2020
Base station 2	5984	7003	–	8838
Base station 3	3078	2020	8838	–

Table 2. Distances between the target points and the base stations in meters

	Base station 0	Base station 1	Base station 2	Base station 3
Target 0	1530	920	7294	2712
Target 1	46	972	6030	3040
Target 2	2572	2944	6043	3427

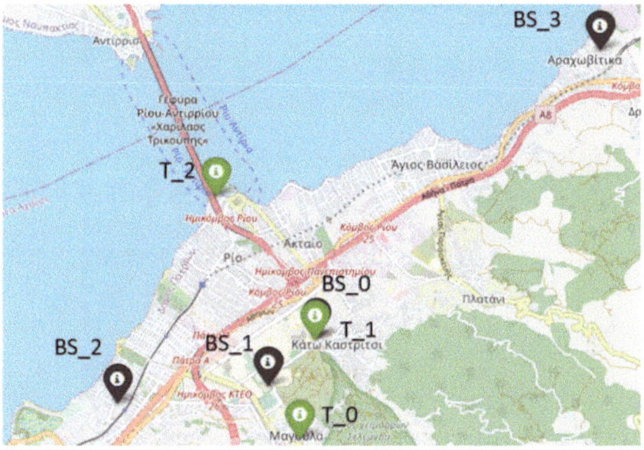

Fig. 5. The locations of the base stations and the targets.

In Fig. 4, the exact locations of the four base stations (BS_0 to BS_3) are illustrated as well as the locations of some targets (T_1, T_2, T_3) that were used in the context of the experiments are presented too (Fig. 5).

Since the position of the target sensor is known, we can calculate the range between the pair of base stations with the reference base station. We repeat the process for every message available. The results are shown in the next figure, where we know that the reference base station should be a specific one, and the Figures illustrate the ranges between the reference base station with other two base stations. In the next Figures, the estimated vs the real distance between each pair of gateways is presented alongside the mean, the std, as well as the maximum and the minimum value (Figs. 6, 7 and 8).

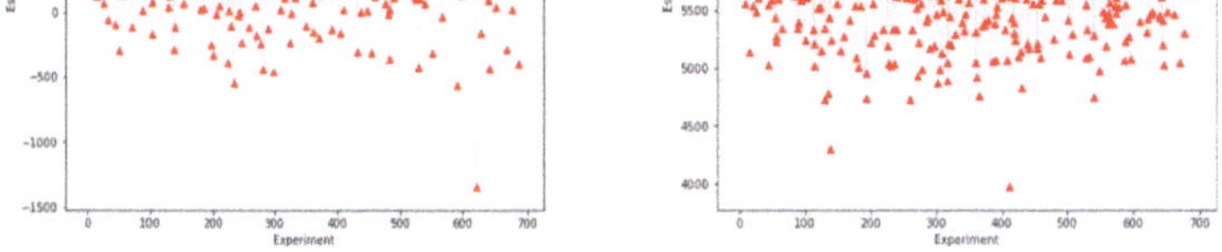

Fig. 6. Real-world data, illustrating the measured vs predicted distances between three gateways from Target_1

As we can observe in the figures, the messages are noisy as individuals and cannot be used as input to any localization algorithm. There are cases where the estimated distance is negative means that due to the noise the actual and the calculated reference base station is different. So when we knew, since the position of the target sensor is known, that the

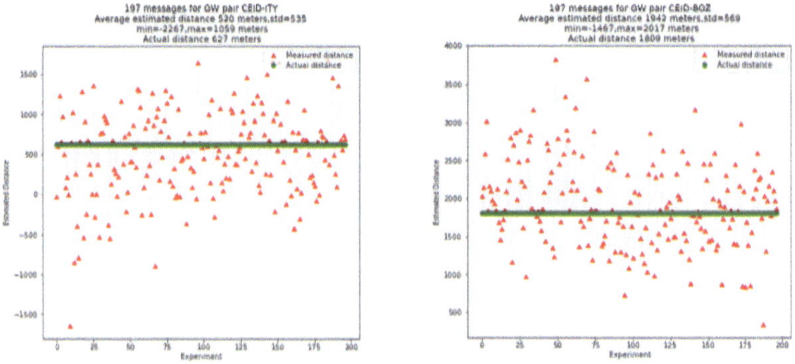

Fig. 7. Real-world data, illustrating the measured vs predicted distances between three gateways from Target_0

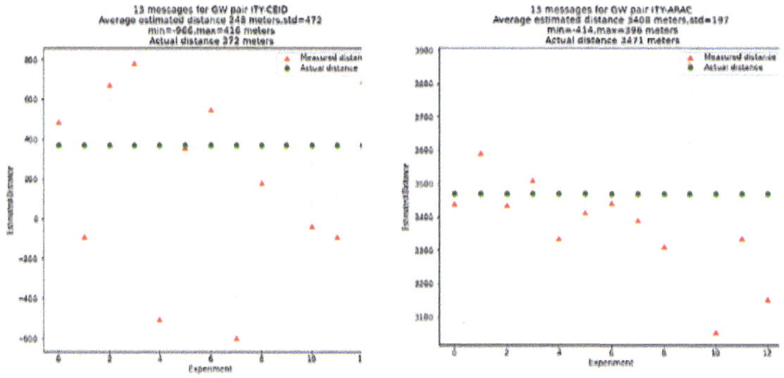

Fig. 8. Real-world data, illustrating the measured vs predicted distances between three gateways from Target_2

reference base station should be the base station BS_0, when we compared the pair BS_0 - BS_1, there are cases that the BS_1 base station received the signal first.

Observing the situation and analyzing other cases with different locations of the sensor node we conclude that to perform localization in LoRa we need to use a group of messages, by the mean that the target sensor will wake up and burst a sequence of messages in a short period. The results collected from the experimental study are illustrated in the next Figures.

In Fig. 9, we present the results of the least-squares algorithm where three base stations were communicating with the wearable device from the location of target 0. In total, 188 groups of messages were formulated and the results of the algorithm in terms of average localization error was 958 m in the context of the 8 × 6 km area having max error of 1500 m and minimum error of 330 m. This is anticipated due to the noisy measurements and the fact that only three base stations communicate with the wearable device.

In Fig. 10 we illustrate the results of the least-squares algorithm where four base stations were communicating with the wearable device. Four groups of messages were

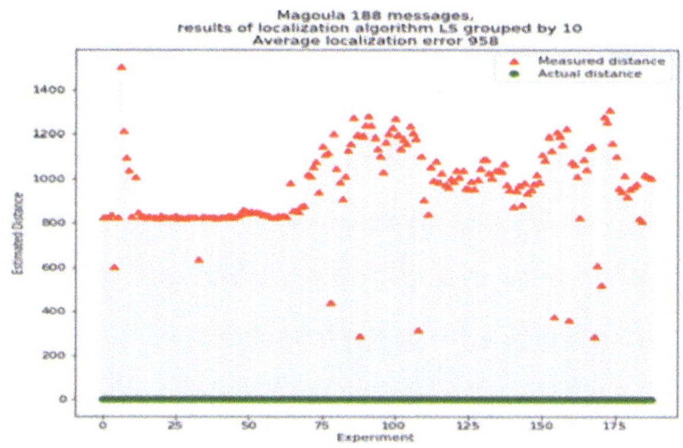

Fig. 9. Least-squares results from target_0

formulated for the location of target 2 and the average localization error was calculated to be 581 m having a very small variance between the maximum and the minimum error. The maximum error was about 620 m and the minimum were about 580 m. We can obtain that the fact that more base stations used in this experiment can reduce the localization error drastically.

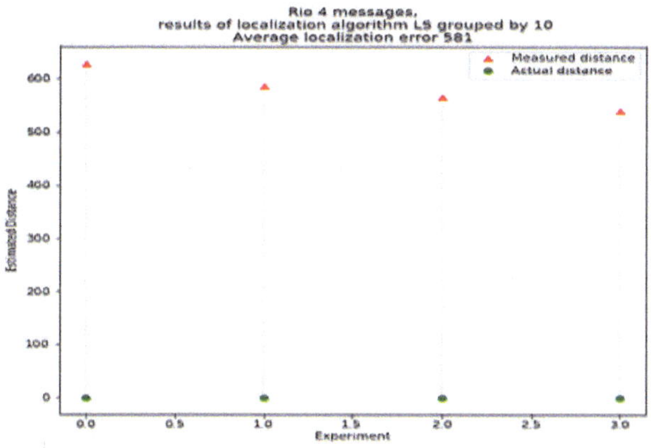

Fig. 10. Least-squares results from target_2

In Fig. 11 the results of Least Squares with the location of target 1 are presented. 684 grouped messages were collected, and the average localization error was 213 m. Once again, all the base stations were able to communicate with the wearable device (i.e. 4 base stations) and the error is drastically reduced, having 500 m maximum error and 45 m minimum.

In Fig. 12, the results of PSO with the location of target 1 are presented. 684 grouped messages were collected and the average localization error was 212 m. Once again, all of the four base stations were able to communicate with the wearable device, and the error is drastically reduced, having 500 m maximum error and 43 m minimum.

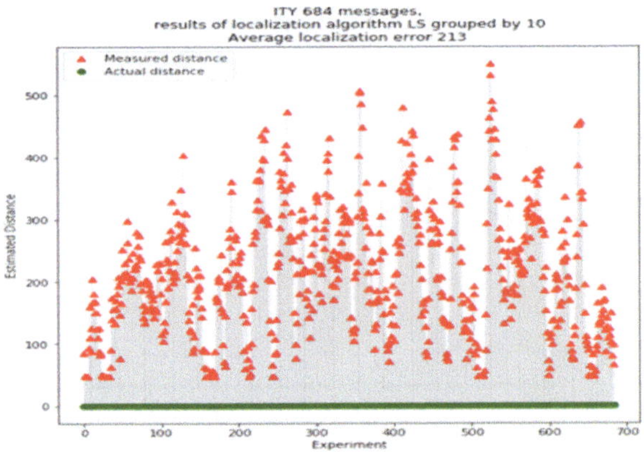

Fig. 11. Least-squares results from target_1

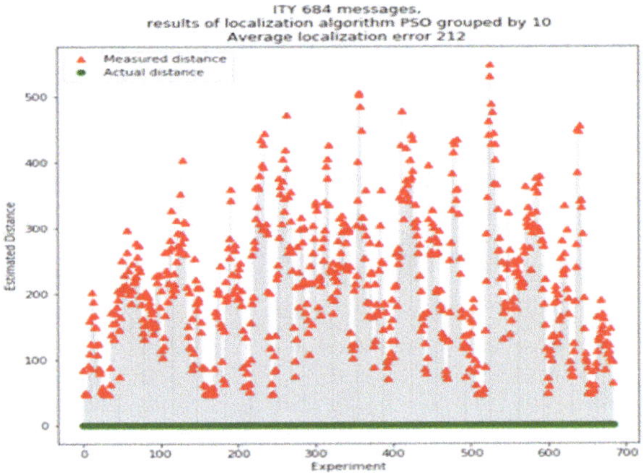

Fig. 12. PSO results from target_1

In Fig. 13, we present the results of the PSO algorithm where three base stations were communicating with the wearable device from the location of target_0. In total, 188 groups of messages were formulated and the results of the algorithm in terms of average localization error was 984 m in the context of the 8 × 6 km area. The maximum error was about 1900 m whilst the minimum error was about 730 m. Same with the least squares we anticipated this error due to the noisy measurements and that only three base stations communicate with the wearable device.

In Fig. 14 we illustrate the results of the PSO algorithm where four base stations were communicating with the wearable device. Thirteen messages were collected resulting in four groups of messages for the location of target 2 and the average localization error using the PSO algorithm was calculated to be 594 m having a very small variance between the maximum and the minimum error. The maximum error was about 620 m and the minimum were about 580 m.

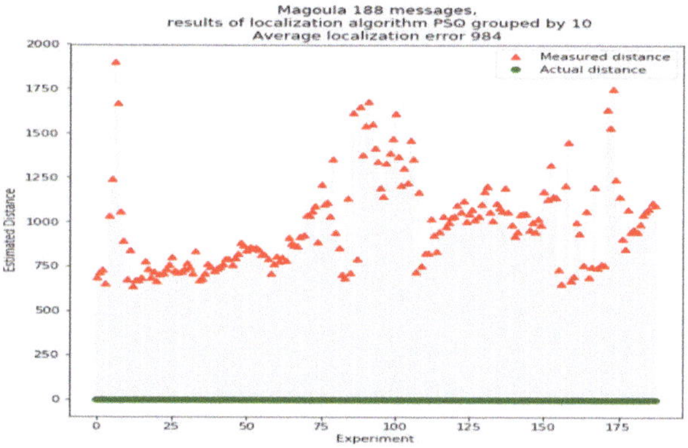

Fig. 13. PSO results from target_0

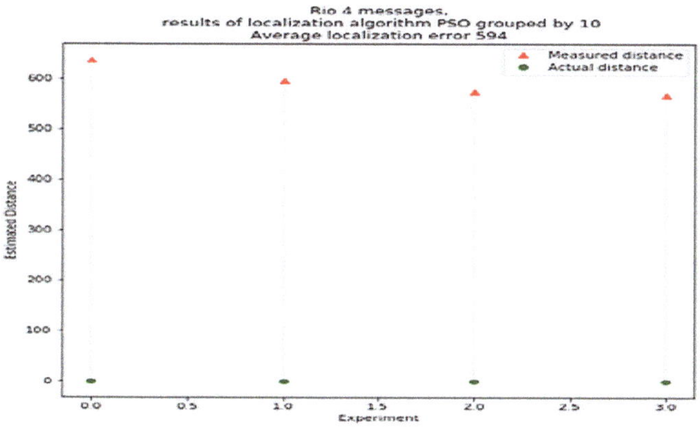

Fig. 14. PSO results from target_2

The results indicate that through LoRa the timestamps of the received messages contain noise, which can result in some cases to ±2 km in the pseudo-ranging distance calculations. Despite the high variance in the prearranging distances, the mean estimated distance is accurate, especially if we consider a big number of messages. The remedy to the noisy measurements was to use a group of ten messages to perform localization. Having a group of ten, reduces drastically the error in pseudo-ranging distances, but still, there might be cases with ±800 m, but the error is reduced at least by 50%. The results indicate that the number of base stations highly affects the localization error. When dealing with three base stations the location estimation might be ambiguous in some cases, also the estimation is affected by the noise in measurements despite the grouping of messages. We can obtain that we three base stations the minimum error was about 300 m and the maximum was about 1800 m while the average localization error through these messages was 958 m. In cases of four base stations, the localization error ranges in 300–700 m, depending on the error in pseudo-ranging distances and the geometry that the target position and the base stations are creating It is obvious that increasing the number of base stations can drastically reduce the localization error by a factor up to 50% when it is combined with the grouping of messages. Regarding the two algorithms we can obtain that they produce similar results regarding the average error but in computational time the least squares can be more effective because the PSO algorithm can scale in computational time if the number of particles and the epochs are close to 200 and higher.

5 Conclusions

In this paper, we examine the localization capabilities of LoRa networks in terms of the localization error. The localization performance of different methods is thoroughly examined under grouping or not of the messages that were sent by the sensor. In the first case, we utilize sets of k messages to calculate the average of TDoAs of these messages and use the average value for every pair of gateways to perform the localization. In the other case, localization is performed by using each one signal sent by the target sensor, and the efficiency and the performance of the algorithms are assessed. Various experiments were performed to assess the localization capabilities of LoRa technology in the real-world setup. The results of the experiments are quite interesting. First, the results indicate that the localization error is greatly affected by the noise of the recorded timestamps of the base stations. The noise results in under and overestimating the pseudo-ranging something that greatly affects the algorithms' performance. Also, the results point out that the grouping of the received messages is quite useful in minimizing the localization error. The main direction for future work concerns the examination of additional localization methods and the assessment of their performance in terms of the localization error. Another direction concerns the design of a bigger scale evaluation on both real data and artificial data that will assist in formulating various scenarios and conditions. These directions constitute the main aspects that future work will examine.

Acknowledgment. This work was partially supported by the Project entitled "Strengthening the Research Activities of the Directorate of the Greek School Network and Network Technologies", funded by the Computer Technology Institute and Press "Diophantus" with project code 0822/001.

References

1. Rauniyar, A., Engelstad, P., Moen, J.: A new distributed localization algorithm using social learning based particle swarm optimization for internet of things. In: IEEE 87th Vehicular Technology Conference (VTC Spring), Porto (2018)
2. Zafari, F., Gkelias, A., Leung, K.K.: A survey of indoor localization systems and technologies. IEEE Commun. Surv. Tutor. **21**(3), 2568–2599 (2019)
3. Daramouskas, I., Kapoulas, V., Pegiazis, T.: A survey of methods for location estimation on Low Power Wide Area Networks. In: 2019 10th International Conference on Information, Intelligence, Systems and Applications (IISA), pp. 1–4. IEEE, July 2019
4. Raychowdhury, A., Pramanik, A.: Survey on LoRa technology: solution for internet of things. In: Intelligent Systems, Technologies and Applications, pp. 259–271 (2020)

5. Marais, J.M., Malekian, R., Abu-Mahfouz, A.M.: LoRa and LoRaWAN testbeds: a review. In: 2017 IEEE Africon, pp. 1496–1501. IEEE, September 2017
6. Saari, M., bin Baharudin, A.M., Sillberg, P., Hyrynsalmi, S., Yan, W.: LoRa—a survey of recent research trends. In: 2018 41st International Convention on Information and Communication Technology, Electronics and Microelectronics (MIPRO), pp. 0872–0877. IEEE, May 2018
7. Gu, C., Jiang, L., Tan, R.: LoRa-based localization: Opportunities and challenges. arXiv preprint arXiv:1812.11481 (2018)
8. Aernouts, M., BniLam, N., Berkvens, R., Weyn, M.: Simulating a combination of TDoA and AoA localization for LoRaWAN. In: Barolli, L., Hellinckx, P., Natwichai, J. (eds.) 3PGCIC 2019. LNNS, vol. 96, pp. 756–765. Springer, Cham (2020). https://doi.org/10.1007/978-3-030-33509-0_71
9. Raza, U., Kulkarni, P., Sooriyabandara, M.: Low power wide area networks: an overview. IEEE Commun. Surv. Tutor. **19**(2), 855–873 (2017). https://doi.org/10.1109/COMST.2017.2652320
10. Kulaib, A.R., Shubair, R.M., Al-Qutayri, M.A., Ng, J.W.: An overview of localization techniques for wireless sensor networks. In: 2011 International Conference on Innovations in Information Technology, pp. 167–172. IEEE, April 2011
11. Podevijn, N., et al.: TDoA-based outdoor positioning with tracking algorithm in a public LoRa network. Wirel. Commun. Mobile Comput. **2018** (2018)
12. Klukas, R., Fattouche, M.: Line-of-sight angle of arrival estimation in the outdoor multipath environment. IEEE Trans. Veh. Technol. **47**(1), 342–351 (1998)
13. Bissett, D.: Analysing TDoA localisation in LoRa networks. Master's thesis, Delft University of Technology, Delft, The Netherlands, October 2018
14. Fargas, B.C., Petersen, M.N.: GPS-free geolocation using LoRa in low-powerWANs. In: Proceedings of the 2017 Global Internet of Things Summit (GIoTS), Geneva, Switzerland, 6–9 June 2017, pp. 1–6 (2017)
15. Aernouts, M., et al.: Combining TDoA and AoA with a particle filter in an outdoor LoRaWAN network. In: 2020 IEEE/ION Position, Location and Navigation Symposium (PLANS), pp. 1060–1069, April 2020
16. Ntseane, L., Isong, B.: Analysis of LoRa/ LoRaWAN challenges. In: 2019 International Multidisciplinary Information Technology and Engineering Conference (IMITEC), pp. 1–7. IEEE, November 2019
17. Lam, K.H., Cheung, C.C., Lee, W.C.: RSSI-based LoRa localization systems for large-scale indoor and outdoor environments. IEEE Trans. Veh. Technol. **68**(12), 11778–11791 (2019)
18. Azmi, N.A., Samsul, S., Yamada, Y., Yakub, M.F.M., Ismail, M.I.M., Dziyauddin, R.A.: A survey of localization using RSSI and TDOA techniques in wireless sensor network: System architecture. In: 2018 2nd International Conference on Telematics and Future Generation Networks (TAFGEN), pp. 131–136. IEEE, July 2018
19. Daramouskas, I., Kapoulas, V., Paraskevas, M.: Using neural networks for RSSI location estimation in LoRa networks. In: Proceedings of the 10th International Conference on Information, Intelligence, Systems and Applications (IISA), PATRAS, Greece, pp. 1–7 (2019). https://doi.org/10.1109/IISA.2019.8900742. 17

Author Index

R. Lee (Ed.): ICIS 2021, SCI 1003, p. 145, 2022.
https://doi.org/10.1007/978-3-030-90528-6

MIX
Papier aus verantwortungsvollen Quellen
Paper from responsible sources
FSC® C105338

If you have any concerns about our products,
you can contact us on
ProductSafety@springernature.com

In case Publisher is established outside the EU,
the EU authorized representative is:
Springer Nature Customer Service Center GmbH
Europaplatz 3, 69115 Heidelberg, Germany

Printed by Libri Plureos GmbH
in Hamburg, Germany